SYSTEMANTICS

SYSTEMANTICS

How Systems Work and Especially How They Fail

JOHN GALL

Illustrated by R. O. Blechman

Quadrangle/The New York Times Book Co.

Third printing, August 1977

First published by Quadrangle/
The New York Times Book Company, Inc., 1977.

Designed by Beth Tondreau.

Previously published by the General Systemantics
Press, Ann Arbor, Michigan.

Library of Congress Cataloging in Publication Data

Gall, John, 1925–
  Systemantics.

  First ed. published in 1975 under title: General
systemantics.
  Bibliography: p.
  1. System theory—Anecdotes, facetiae, satire, etc.
I. Title.
PN6231.S93G   1977   003   76–50820
ISBN 0–8129–0674–8

CONTENTS

PREFACE: HOW THIS TREATISE CAME TO BE WRITTEN

The immediate motivation for undertaking this work was provided by the bizarre experience of a colleague of the author (let us call him Jones), a medical scientist specializing in the study of mental retardation. This field, until recently, was a very unfashionable one, and Jones considered himself fortunate to be employed as Research Associate at a small State Home for retarded children. In this humble, even despised, position he was too low on the Civil Service scale to merit the attention of administrators, and he was therefore left alone to tinker with ideas in his chosen field. He was happily pursuing his own research interests when, following presidential interest and national publicity, mental retardation suddenly became a fashionable subject. Jones received an urgent invitation to join an ambitious federally funded project for a systematic attack upon the "problem"* of mental retardation.

Thinking that this new job, with its ample funds and facilities, would advance both his research efforts and his career, Jones joined. Within three months his own research had come to a halt, and within a year he was

*See The "Problem" Problem, Chapter XI.

completely unable to speak or think intelligently in the field of mental retardation. He had, in fact, become *retarded* relative to his previous condition.

Looking about him to discover, if possible, what had happened, Jones found that a dozen other skilled professionals who made up the staff of the project had experienced the same catastrophe. That ambitious project, designed to advance solutions to the "problem," had, in fact, taken most of the available workers in the field and *neutralized* them.

What had gone wrong? Jones, knowing of the author's interest in systems operation, turned to him for advice, and the two of us, jointly, tried to analyze the problem. We first of all reviewed Parkinson's classic essay on Institutional Paralysis,* hoping to find enlightenment there. Was the disease a fulminating case of Injelititis?** Obviously not. The professional staff of the Institute were competent, dedicated, and hardworking. Furthermore, the administrators were experienced, energetic, and extremely logical in their approach to problems. The Institute was *failing in spite of the best efforts of every one of its members.*

Was it an example of the Peter Principle,*** in which members had ascended the hierarchy until they had reached jobs for which they were not fitted? No. This was a new organization, and no one had been promoted very far. *Slowly it began to dawn on us that men do not yet*

---

*C. Northcote Parkinson. *Parkinson's Law and Other Studies in Administration.* Boston: Houghton Mifflin, 1957.
**Ibid. Injelititis: Incompetence and jealousy interacting according to the formula I^3J^5.
***Laurence J. Peter and Raymond Hull. *The Peter Principle.* New York: Bantam Books, 1970.

understand the basic laws governing the behavior of complex organizations. A great enterprise can flounder helplessly or even wither away before our very eyes as the result of malignant but as yet unnamed disorders, or in response to the operation of natural laws whose Newton has not yet been born to announce them to mankind.

Faced with this realization, and moved by the dramatic and touching crisis that had overtaken his colleague, the author resolved to redouble his researches into the causes of organizational ineptitude and systems malfunction, seeking deep beneath the surface for the hidden forces that cause the best-laid plans to "gang aft agley." Little did he suspect, at that moment, where those studies would lead. He had not yet experienced the blinding illumination of the OPERATIONAL FALLACY. The FUNDAMENTAL THEOREM OF SYSTEMANTICS lay well over the horizon. Even the relatively simple and easy-to-understand GENERALIZED UNCERTAINTY PRINCIPLE was not yet a gleam. Those and other deep-ranging and depressing generalizations were to come only much later, after exhaustive researches conducted under the least auspicious circumstances and in settings not calculated to bolster the faint of heart.

What follows is the fruit of those researches set forth as briefly and simply as possible, in a style that is deliberately austere and (the author may be permitted to hope) not without a certain elegance, derived from the essentially mathematical nature of the science itself. The reader must imagine for himself at what cost in blood, sweat, and tears —and in time spent in deep contemplation of contemporary systems—these simple statements have been wrung from the messy complexity of the real world. They are offered in the hope and faith that a *knowledge of the natural laws of complex systems* will enable mankind to avoid some of the more egregious errors of the past.

At the very least this little book may serve as a warning to those who read it, thus helping to counter the headlong rush into Systemism* that characterizes our age. And who knows? Perhaps readers of this modest treatise will be stimulated to discover new Systems-insights of their own, that could lead to even higher achievements for the infant science of Systemantics.

---

*Systemism: n.1. The state of mindless belief in systems; the belief that systems can be made to function to achieve desired goals. 2. The state of being immersed in systems; the state of being a Systems-person. (See Chapter VI: "Inside Systems.")

INTRODUCTION: PARADOX LOST AND FOUND. DIMENSIONS OF THE PROBLEM

All around us we see a world of paradox: deep, ironic, and intractable. A world in which the hungry nations export food; the richest nations slip into demoralizing economic recessions; the strongest nations go to war against the smallest and weakest and are unable to win; a world in which revolutions against tyrannical systems themselves become tyrannies. In human affairs, celebrities receive still more publicity because they are "well known"; men rise to high positions because of their knowledge of affairs only to find themselves cut off from the sources of their knowledge; scientists opposed to the use of scientific knowledge in warfare find themselves advising the government on how to win wars by using scientific knowledge . . . the list is endless. Ours is a world of paradox.

Why is this? How does it come about that things turn out so differently from what common sense would expect?

The religious person may blame it on original sin. The historian may cite the force of trends such as population growth and industrialization. The sociologist offers reasons rooted in the peculiarities of human associations. Reformers blame it all on "the system" and propose new systems that would, they assert, guarantee a brave new world of justice, peace, and abundance. Everyone, it seems, has his own idea of what the problem is and how it can be corrected. But all agree on one point—that their own system would work very well if only it were universally adopted.

The point of view espoused in this essay is more radical and at the same time more pessimistic. Stated as succinctly as possible: the fundamental problem does not lie in any particular system but rather in *systems as such*. Salvation, if it is attainable at all, even partially, is to be sought in a deeper understanding of the ways of systems, not simply in a criticism of the errors of a particular system.

But although men build systems almost instinctively,* they do not lightly turn their ingenuity to the study of How Systems Work. That branch of knowledge is not congenial to man, it goes against the grain. Goal-oriented man, the upright ape with the spear, is interested in the *end-result*. If the spear flies wide of the mark, man is equally likely to trample it to bits in a rage or to blame the erratic flight on malevolent spirits. He is much less likely to undertake a critical analysis of hand-propelled missiles, and infinitely less likely to ponder the austere abstractions presented in this book.

If young people lack experience and interest for understanding How Systems Work, older people are already defeated. They may have learned from direct experience a few things about systems, but their experience will have been fragmentary and painful—and in any case, for them the battle is over. No, only a handful—only a lucky few

---

*Recent research has linked this impulse to nesting behavior in birds and to token-collecting in higher primates.

—ever come to clear awareness of this dread and obscure subject.

No one, these days, can avoid contact with systems. Systems are everywhere: big systems, little systems, systems mechanical and electronic, and those special systems that consist of organized associations of people. In self-defense, we must learn to live with systems, to control *them* lest *they* control *us*. As Humpty Dumpty said to Alice (though in another context): "The question is: which is to be master—that's all."

No one can afford not to understand the basic principles of How Systems Work. Ignorance of those basic laws is bound to lead to unrealistic expectations of the type that have plagued dreamers, schemers, and so-called men of affairs from the earliest times. Clearly there is a great need for more widespread knowledge of those basic laws. But (and just here is another example of the paradoxical nature of systems-functions) there is a strange dearth of available information written for the general reader. Technical tomes of systems analysis and operations research abound on the shelves of science libraries and of business management institutes. But nowhere is there to be found a single, basic primer that spells out the essential pragmatic facts of systems in the form of simple and easy-to-grasp axioms. Similarly there are no courses in Systems Function in our high schools or junior colleges. Like sex education, systems sophistication has until recently been a *taboo subject*. This book breaks the taboo. It tells all, in frank and intimate language understandable to anyone who is willing to read and reflect. No longer can people take refuge in the plaint, "Nobody told me." It's all here, within the covers of one small book.

SYSTEMANTICS

I. HISTORICAL OVERVIEW

All over the world, in great metropolitan centers as well as in the remotest rural backwaters, in sophisticated electronics laboratories, and in dingy clerical offices, people everywhere are struggling with a Problem:

Things Aren't Working Very Well. *

This, of course, is nothing new. People have been discouraged about things in general many times in the past. A good deal of discouragement prevailed during the Dark Ages, and morale was rather low in the Middle Ages, too. The Industrial Revolution brought with it depressing times, and the Victorian Era was felt by many to be particularly gloomy. The Atomic Age isn't remarkable for cheer, either. At all times there have been people who felt that things weren't working out very well. This observation has gradually come to be recognized as an ongoing fact of life, an inseparable component of the human condition. Because of its central role in all that follows (being the fundamental observation upon which all further re-

*For an extensive review of things that aren't working very well at present, see Peter and Hull, *op. cit.,* Introduction, pp. ix-xviii.

search into systems has been based) it is known as The
Primal Scenario. We give it here in full:

> Things (Things Generally / All Things / The
> Whole Works) Are Indeed Not Working Very
> Well. In Fact, They Never Did.

In formal systems terminology it may be stated con-
cisely in axiomatic form:

> Systems In General Work Poorly Or Not At All.

But this fact, repeatedly observed by men and women
down through the ages, has been, in the past, always attributed to various special circumstances. It has been reserved for our own time, and for a small band of men of
genius, working mostly alone, to throw upon the whole
subject the brilliant light of intuition, illuminating for all
mankind the previously obscure reasons why Things So
Often Go Wrong, or Don't Work, or Work in Ways Never
Anticipated. To list the names of those contributors is to
recite the Honor Roll of Systemantics.

No history of the subject would be complete without
some reference to the semilegendary, almost anonymous
Murphy (*floreat circa* 1940?) who chose to disguise his
genius by stating a fundamental systems theorem in com-
monplace, almost pedestrian terminology. This law,
known to schoolboys the world over as *Jellybread always
falls jelly-side down,* is here restated in Murphy's own
words, as it appears on the walls of most of the world's
scientific laboratories:

If Anything Can Go Wrong, It Will.

In the Law as thus formulated, there is a gratuitous and unjustified element of teleology, an intrusion of superstition, or even of belief in magic, which we today would resolutely reject. The Universe is not actually malignant, it only *seems* so.*

Shortly after Murphy there appeared upon the scene a new and powerful mind, that of Count Alfred Korzybski, in whose honor the entire field of General Systemantics has been named. Korzybski was the author of *General Semantics,* a vaulting effort at a comprehensive explanation of Why Things Don't Work. This early attempt to pinpoint the flaw in human systems was itself flawed, however, by the author's monistic viewpoint. Korzybski had convinced himself that all breakdowns of human systems are attributable to misunderstandings—to failures of communication.

Korzybski failed to grasp the essential point that human systems are not prevented from working by some single, hidden defect, whether of communication or anything else. Failure to function as expected (as we shall show later) is an *intrinsic feature* of systems, resulting from laws of systems-behavior that are as rigorous as any in Natural Science or Mathematics. Hence, the appropriateness of the term GENERAL SYSTEMANTICS for the entire field. It is a perfectly general feature of systems not to do what they are intended to do. Furthermore, the word ANTICS hidden in the term carries this implication in a

*See "The Mysterious Ways of Systems," Chapter III.

lively way. SYSTEMS DISPLAY ANTICS. They "act up." Nevertheless, as we shall see, Korzybski, by stressing the importance of precise definitions, laid the groundwork for the Operational Fallacy, which is the key to understanding the paradoxical behavior of systems (see Chapter V).

After Korzybski, a brilliant trio of founders established the real basis of the field. Of these, the earliest was Stephen Potter,* who painstakingly elaborated a variety of elegant methods for bending recalcitrant systems to the needs of personal advancement. It must be admitted that Potter was essentially a pragmatist whose goals were utilitarian and whose formulations lack the broad generalities of a Parkinson or a Peter.

Following Potter, C. Northcote Parkinson established an undying claim to fame by prophesying—as early as 1957—the future emergence of the problem of Table Shape in diplomatic conferences.** He was, of course, triumphantly vindicated in the Paris Peace Talks of 1968, when an entire season was devoted to just this topic before discussion of the cessation of hostilities could even begin. No clearer demonstration of the Generalized Uncertainty Principle could have been asked.***

Third in the brilliant trio of founders is Doctor Laurence J. Peter, whose Principle of Incompetence comes close to being the Central Theorem of Administrative Systemantics.

---

*Stephen Potter. *One-upmanship.* New York: Henry Holt & Co., 1952.
**Parkinson, *op. cit.,* p. 17.
***True, Parkinson did not recognize it as such at the time.

Having paid tribute to all these men, however, one must recognize that the infant science on whose foundations these giants were working (one must mix metaphors now and then) was still limited. There was no organized set of basic principles from which to operate. The foundations had been laid erratically, a piece at a time, by individual workers of genius.

Still needed was a systematic exposition of the fundamental principles—the axioms—upon which all later superstructures could be built.

The present work is humbly offered as a first approach to that goal. It will have its shortcomings, of course. The individual propositions will be argued, dissected, and criticized; and then either rejected as trivial, erroneous, or

incomprehensible; or enshrined in the literature as having stood the test of open debate and criticism. This is as the author would wish it.

In the pages that follow, we shall be principally concerned with systems that involve human beings, particularly those very large systems such as national governments, nations themselves, religions, the railway system, the post office, the university system, the public school system, etc., etc., etc. But in our formulations of the laws of such systems, we have striven for the greatest possible degree of generality. If we are correct, our theorems apply to the steamship itself as well as to the crew who run it and to the company that built it.

Here, then, is the very first Book of Systems Axioms, the very first attempt to deal with the cussedness of Systems in a fundamental, logical way, by getting at the basic rules of their behavior.

II. FIRST PRINCIPLES

We begin at the beginning, with the Fundamental Theorem:

New Systems Mean New Problems.

Explanation: When a system* is set up to accomplish some goal, a new entity has come into being—the system itself. No matter what the "goal" of the system, it immediately begins to exhibit system behavior; that is, to act according to the general laws that govern the operation of all systems. Now the system itself has to be dealt with. Whereas before, there was only the problem—such as warfare between nations, or garbage collection—there is now an additional universe of problems associated with the functioning or merely the presence of the new system.

In the case of garbage collection, the original problem could be stated briefly as: "What do we do with all this garbage?" After setting up a garbage-collection system, we find ourselves faced with a new universe of problems. These include questions of collective bargaining with the garbage collectors' union, rates and hours, collection on very cold or rainy days, purchase and maintenance of garbage trucks, millage and bond issues, voter apathy, regulations regarding separation of garbage from trash, etc., etc.

*For definition of system, see Appendix VI.

Although each of these problems, considered individually, seems to be only a specific technical difficulty having to do with setting up and operating a garbage-collecting system, we intend to show that such problems are really specific examples of the operation of general laws applicable to any system, not just to garbage collecting. For example, absenteeism, broken-down trucks, late collections, and inadequate funds for operation are specific examples of the general Law that *LARGE SYSTEMS USUALLY OPERATE IN FAILURE MODE.* Again, if the collection men bargain for more and more restrictive definitions of garbage, refusing to pick up twigs, trash, old lamps, etc., and even leaving behind properly wrapped garbage if it is not placed within a regulation can, so that eventually most taxpayers revert to clandestine dumping along the highway, this exemplifies the *Le Chatelier's Principle** (THE SYSTEM TENDS TO OPPOSE ITS OWN PROPER FUNCTION), a basic law of very general application. These and other basic laws of systems function are the subject of subsequent chapters.

In most towns of small-to-medium-size, a garbage-collecting system qualifies as a small-to-medium-sized system, and systems of such size often do accomplish a measurable fraction of what they set out to do. *Some garbage does get collected.* The original problem is thereby somewhat reduced in magnitude and intensity. Over against this benefit, however, one must balance the new problems facing the community, the problems of administering and maintaining the collection system.

*to come

The sum total of problems facing the community has not changed. They have merely changed their form and relative importance. We require at this point a Definition:

> Any state or condition of the Universe, or of any portion of it, that requires the expenditure of human effort or ingenuity to bring it into line with human desires, needs, or pleasures is defined as an ANER-GY-STATE.
> ANERGY is measured in units of effort required to bring about the desired change.

Now we are in position to state a Theorem of sweeping generality:

> The Total Amount Of Anergy In The Universe Is
> Fixed.

This is known, naturally, as the Law of Conservation of Anergy.

We offer without proof the following Corollary:

> Systems Operate By Redistributing Anergy Into
> Different Forms And Into Accumulations Of
> Different Sizes.

One school of mathematically oriented systems theoreticians holds that the Law of Conservation of Anergy is only approximately true. According to them, in very large systems a Relativistic Shift occurs, whereby the total amount of Anergy increases exponentially. In really large and ambitious systems, the original problem may persist unchanged and at the same time a multitude of new problems arise to fester (or ferment) unresolved. The Relativists point to garbage collection in large metropolitan areas as an example. Not only does the garbage not get collected, but also armies of striking workers must be fed and clothed, the multitudes of the city must be immunized against diseases of filth, the transportation systems break down because cars and buses cannot get around the mountains of refuse, and things in general quickly go to an extreme degree of disrepair. Granted that some of these effects would have been present to some degree had there never been any garbage-collection system at all, it is clear that they become *much worse* because people had come to rely on the system.

Laws of Growth.

Systems are like babies: once you get one, you have it. They don't go away. On the contrary, they display the most remarkable persistence. They not only persist; they grow. And as they grow, they encroach. The growth potential of systems was explored in a tentative, preliminary way by Parkinson, who concluded that administrative systems maintain an average rate of growth of 5 to 6 percent per annum regardless of the work to be done. Parkinson was right so far as he goes, and we must give him full honors for initiating the serious study of this important topic. But what Parkinson failed to perceive, we now enunciate—the general systems analogue of Parkinson's Law.

The System Itself Tends To Grow At 5 To 6
Percent Per Annum.

Again, this Law is but the preliminary to the most
general possible formulation, the Big-Bang Theorem of
Systems Cosmology:

Systems Tend To Expand To Fill The Known
Universe.*

We have remarked that systems not only grow, they
also encroach. An entire volume could be devoted to researches in this area alone. Innumerable examples of the
phenomenon of encroachment can be found everywhere in
our society. A striking example is the "do-it-yourself"
movement, instigated by managers of the largest and most
sophisticated system of mass production in the world to
*make the consumer do some of the work the system is
supposed to do.* The consumer is encouraged to do the
work of assembling the parts of the product he has bought
on the bizarre grounds that "it saves so much work and
expense." Several hours later, the exasperated and frustrated purchaser may recall that the system of mass production was set up in the first place because such functions
can be more cheaply and rapidly done under factory conditions. The system simply prefers to encroach; that is, to
make someone else do the work.

Pushing the expenses off on the consumer goes back at
least as far as the *ancien régime* in France, where the

---

*That this outcome does not occur in fact is due to the existence
of various inhibitory forces (see Chapter IV).

peasants were subjected to grinding taxation to support the aristocracy, who were not taxed.* The system of government, at its basis a system for protecting the people, encroached upon them until it became their worst oppressor. In the United States, the Internal Revenue Service not only collect our taxes, they also make us compute the tax for them, an activity that exacts an incalculable cost in sweat, tears, and agony and takes years off our lives as we groan over their complicated forms.

In accordance with the same principle, patients in hospitals are blamed for not getting well or for reacting badly to medicine or surgery. Motorists whose cars develop engine trouble are ticketed by the police, or their vehicles are towed away and crushed into scrap metal. And schoolboys who have trouble learning their lessons are punished by their teachers. Truly, Systems Expand, and as they expand, they Encroach.

*A similar tendency can be discerned in the tax policies of the United States. It reached its climax in the late 1960s with the practice among high government officials of writing-off one's official papers at inflated valuations. The practice has been designated (by us) as *Nixation*, or negative (nix-) taxation for the well-to-do.

III. THE MYSTERIOUS WAYS OF SYSTEMS

The Primal Scenario enshrines the universal observation that Things Don't Work Very Well. For most people, it is enough to sigh and remark that Things Were Ever Thus. However, to the serious systems-student, a world of absorbing interest lies beneath the superficial observation of things as they are. The more one delves beneath the surface—the more one becomes committed to close observation and, most importantly, to generalization from particular experience—the more one begins to attain insight into the true and deeper meanings of things. Such is the case with the Primal Scenario. Let us ask what lies beneath the surface of that melancholy but universal observation. Just what is the characteristic feature of things not working out? To ask the question is to see the answer, almost immediately. It is the element of *paradox,* to which we have already alluded. Things not only don't work out well, they work out in strange, even paradoxical, ways. Our plans not only go awry, they produce results we never expected. Indeed they often produce the opposite result from the one intended.

Item: Insecticides, introduced to control disease and improve crop yields, turn up in the fat pads of Auks in the Antipodes and in the eggs of Ospreys in the Orkneys, resulting in incalculable ecologic damage.

Item: The Aswan Dam, built at enormous expense to improve the lot of the Egyptian peasant, has caused the Nile to deposit its fertilizing sediment in Lake Nasser, where it is unavailable. Egyptian fields must now be artificially fertilized. Gigantic fertilizer plants have been built to meet the new need. The plants require enormous amounts of electricity. The dam must operate at capacity merely to supply the increased need for electricity which was created by the building of the dam.

Item: Many backward nations, whose greatest need is food to feed their people, sell their crops and bankrupt themselves to buy—not food—but advanced military hardware for the purpose of defending themselves against their equally backward neighbors, who are doing the same thing.

Examples could be multiplied indefinitely, and the reader is encouraged to provide his own, as an exercise.

What is the common element in all these surprising and paradoxical situations? It is just this: in each case *a complex system has failed to act as expected by its designers,* but has instead exhibited behavior that no one expected it to exhibit.

Now, so long as one is content merely to make the observation that a particular system isn't working well, or isn't doing what is expected, one is really only at the level of insight summarized in the Primal Scenario. The crucial step forward in logic is a small one, but it is of critical importance for progress in Systems-thinking. There is a world of difference, psychologically speaking, between the passive observation that Things Don't Work Out Very Well, and the active, penetrating insight that

Complex Systems Exhibit Unexpected Behavior.

The one is merely a pessimistic feeling; the other conveys the exhilarating euphoria that accompanies the recognition of a Law of Nature.

Incredibly enough, the first big breakthrough in recognition of the GENERALIZED UNCERTAINTY PRINCIPLE (G.U.P.) did not come until the 1950s, when a daring—if anonymous—group of biologists toppled Watsonian determinism with one short, pithy aphorism now known as the Harvard Law of Animal Behavior:

> Under Precisely Controlled Experimental
> Conditions, A Test Animal Will Behave As It
> Damn Well Pleases.

The formulators of this Law failed to generalize to systems as such, thereby missing—by a whisker, so to speak—their chance of immortality:

> *Not Just Animal Behavior, But The Behavior Of*
> *Complex Systems Generally, Whether Living Or*
> *Nonliving, Is Unpredictable.*

It is fitting that this Law should have been foreshadowed, even if in limited form, by biologists, for they, more than others, are brought face to face in their daily professional activities with the essential unpredictability of living things. Mathematicians and engineers have considerably more difficulty with the G.U.P. Accustomed as they are to creating systems out of their own heads, they are affronted that such systems—their own creatures, so to speak—should exhibit behavior that was unplanned and even undreamed of by themselves. Some go so far as to assert that the G.U.P. is "not logical."

We sympathize with that point of view, or rather, with the visceral reaction underlying it. As long as a system exists only in the head of its creator, we would agree that it might be knowable in all its implications. But once that system is translated into the real world, into hardware and people, it becomes something else. It becomes a real-world thing, and mere mortals can never know all there is to know about the real world. Therein lies the inevitability of the G.U.P.

The operation of the G.U.P. is perhaps most clearly displayed in the realm of Climax Design, i.e., in the construction of the largest and most complex examples of man-made systems, whether buildings, ships and planes, or organizations. The ultimate model (the largest, fastest, tallest, etc.) often, if not invariably, exhibits behavior so unexpected as to verge on the uncanny. The behavior is often an unsuspected way of failing, to which, because of its importance, we have devoted an entire chapter.* Let us review only a few examples of Climax Design:

Item: The largest building in the world, the space vehicle preparation shed at Cape Kennedy, *generates its own weather, including clouds and rain.* Designed to protect space rockets from the elements, it pelts them with storms of its own.

Item: The Queen Elizabeth II, greatest ocean liner ever built, has three separate sets of boilers for safety, reliability, and speed. Yet on a recent cruise, in fine weather and a calm sea, all three sets of boilers *failed simultaneously.*

*Chapter IX, Systems-Failure (Theory of Errors).

Item: More disquieting, if no surprise to the serious Systems-student, is the fact that the Bell Telephone System, largest in the world, is beginning to exhibit cracks in the System—vagaries of behavior that not even the company experts can entirely explain. The crucial change probably began about the time when the operator's familiar request, "Number, please" suddenly reversed its meaning. No longer was the young lady at Central asking you to tell her the number you wanted to reach; she was asking for *your own number,* for purposes beyond a layman's comprehension. Since that moment, the changes have come with increasing rapidity. People are discouraged from dialing the operator and are made to feel guilty if they do. But more to the point: the sophisticated computers that reroute your telephone call through five states in order to utilize the least-congested circuits have begun putting you through California when you call next door.

You may lift the receiver to dial a number and discover you are eavesdropping on a transatlantic conversation; you may dial your party and end up in a three- or four-way conversation with a neighbor and several perfect strangers. The end is not yet in sight.*

We formalize these observations in the *Non-Additivity Theorem* of Systems-behavior, alternatively known as the Climax Design Theorem:

> A Large System, Produced By Expanding The
> Dimensions Of A Smaller System, Does Not
> Behave Like The Smaller System.

*It is conceivable that private citizens will begin bugging their own phones in order to learn what is going on. In this practice their leaders are clearly well ahead of the public.

IV. FEEDBACK

Every student of science is required at some point or other in his career to learn Le Chatelier's Principle. Briefly, this Law states that any natural process, whether physical or chemical, tends to set up conditions opposing the further operation of the process. Although the Law has very broad application, it is usually looked upon more as a curiosity than as a profound insight into the nature of the Universe. We, on the other hand, regard it as a cornerstone of General Systemantics. No one who has had any experience of the operation of large systems can fail to appreciate its force, especially when it is stated in cogent Systems-terminology as follows:

> Systems Get In The Way.

Or alternatively:

> The System Always Kicks Back.

In slightly more elegant language:

> Systems Tend To Oppose Their Own Proper
> Functions.

In the field of human organizations, probably the outstanding example of Le Chatelier's Principle occurs in connection with the Goals and Objectives Mania, a specialized manifestation of a very ancient and widespread phenomenon known as Administrative Encirclement.*

Let us take as an example the case of Lionel Trillium, a young Assistant Professor in the Department of Botany at Hollyoak College. Although it is now too late to do any good for poor Trillium, we shall review his case history, step by step, in the hope that similar tragedies may be averted in the future.

Trillium's Department Head, Baneberry, has for some years now failed to initiate new and interesting hypotheses about the behavior of the Slime Molds, his chosen area of specialization. Paralleling this decline of scientific productivity, he has exhibited increasing interest in improving the "efficiency" of his Department. (The medically oriented reader will recognize in these symptoms the insidious onset of intellectual menopause.) Baneberry has actually gone to the extreme of checking out of the library some recent publications on management science. Before his jaded eyes a new world has been revealed, and his mind is now buzzing with the terminology of Information Retrieval Systems, Technology Assessment, Program Budgeting, and above all, Management by Goals and Objectives. He fires off a memo to the staff of his Department requiring them to submit to him, in triplicate, by Monday next, statements of their Goals and Objectives.

This demand catches Trillium at a bad time. His studies of angiosperms are at a critical point. Nevertheless, he must take time out to consider his Goals and Objectives, as the wording of the memo leaves little doubt of the

*For administrators' neuroses, see Chapter VI.

consequences of failure to comply.

Now Trillium really does have some personal goals of his own, that his study of Botany can advance. In actual fact, he entered that field because of unanswered questions* having to do with the origin of babies. His boyhood curiosity was never satisfied, and it became fixed on the mechanics of reproductive processes in living creatures. But to study reproduction directly, in animals, creates too much anxiety. Therefore he has chosen the flowering plants, whose blatant sexuality is safely isolated from our own. Trillium is happy as a botanist, and never so happy as when he is elucidating the life cycle of an angiosperm.

But now his Chief is demanding Goals and Objectives. This is both disturbing and threatening. Trillium doesn't want to think about his real goals and objectives; indeed, they are unknown to his conscious mind. He only knows he likes Botany.

But he can't just reply in one line, "I like botany and want to keep on studying it." No, indeed! What is expected is a good deal more formal, more organized, than that. It should fill at least three typewritten sheets, single-spaced, and should list Objectives and Subobjectives in order of priority, each being justified in relation to the Overall Goal and having appended a time-frame for completion and some criteria for determining whether they have been achieved. Ideally, each paragraph should contain at least one reference to DNA or the phrase "double helix." Trillium goes into a depression just thinking about it.

Furthermore, he cannot afford to state his true goals.

*That he never dared ask his parents.

He must at all costs avoid giving the impression of an ineffective putterer or a dilettante. He must not appear fuzzy-headed. His goals must be well-defined, crisply stated, and must appear to lead somewhere important. They must imply activity in areas that tend to throw reflected glory on the Department. After all, one can't expect a University to maintain a Department just for people who have a faintly scurrilous interest in how plants reproduce. Therefore, Trillium is forced to include in his statement all kinds of things that he's really not the least interested in.

Trillium struggles with all these considerations, which he does not consciously formulate. He only feels them as a deep malaise and sense of confusion that seizes him whenever he thinks about writing the Goals and Objectives statement. He puts it off as long as possible, but still it interferes with his studies of angiosperms. He can't concentrate. Finally he gives up his research, stays home three days, and writes the damned thing.

But now he is committed in writing to a program, in terms of which his "success" can be objectively assessed by his Chief. If he states that one objective for the coming year is to write three papers on angiosperms and he actually writes only two, he is only 67 percent "successful," even if each of the two papers is a substantial contribution to his field. Or he may not write any papers at all, devoting the time instead to a book for which the idea has unexpectedly matured. In that case his "success" is zero. No matter that he has overachieved in an unexpected area. His failure to achieve his stated objectives is demonstrable in black and white. Trillium has been led, step by step, down the primrose path of logic to disaster. The logic was devised by others, but that is no consolation to him.

The next step is even more catastrophic. Because Trillium has clearly stated his Goals and Objectives, it is now possible to deduce with rigorous logic how he should spend his waking and working hours in order to achieve them most efficiently. No more pottering around pursuing spontaneous impulses and temporary enthusiasms! No more happy hours in the Departmental greenhouse! Just as a straight line is the shortest distance between two points, so an efficient worker will move from Subobjective A to Subobjective B in logical pursuit of Objective K, which leads in turn toward the Overall Goal. Trillium can be graded, not only on his achievements for the year, but also on the efficiency with which he moves toward each objective. He has become *administratively encircled.* The administrators, whose original purpose was to keep track of writing supplies for the professors, now have the upper hand and sit in judgment on their former masters.

Only one step remains to complete Trillium's shackling in the chains he himself has helped to forge. On advice of the University administrators, the legislators of his State establish by law the number of hours a Professor of Botany must spend on each phase of his professional activities. Trillium may feel impelled to protest, but how can he? The lawmakers are only formalizing what he himself has told them, through his Goals and Objectives statements, he wants to do! The System of Management by Goals and Objectives, designed to improve Trillium's efficiency and measure his performance as a botanist, has *gotten in the way, kicked back,* and *opposed its own proper function.* Once more the universal validity of Le Chatelier's Principle has been demonstrated.

Baneberry, meanwhile, has been powerfully reinforced in his new-found function as judge of other botanists. His own botanical career may be wilting, but he has found a new career in assessing the strengths and weaknesses of his colleagues—especially their weaknesses. The heady experience of hobnobbing with legislators and people of power in the "real" world gives him a new lease on life and convinces him that Trillium really is just a poor potterer who will never amount to anything. If only Botany could attract *men of action* like the lawgivers with whom he has been wining and dining!*

The Power of Positive Feedback—A Warning.

We have seen that the natural tendency of systems is to set up negative feedback to their own operations (Le Chatelier's Principle). But what about *positive feedback?*

A great deal of nonsense is talked these days about positive feedback. When the term is used merely as an inflated way of referring to praise given a person for a job well done, no particular harm results. But feedback in electronic systems leads to oscillation, resulting in a loud squeal and loss of function. The entire energy of the system suddenly goes down the drain in a burst of useless or even harmful noise. In mechanics, the ill-fated Electra airplane was a victim of positive feedback of energy from the propellers to the wings. Political rallies, with their engineered positive feedback, provide the same ominous sense of things vibrating out of control. No, students of Systemantics, positive feedback into a large system is a

---

*Administrator's Grandiosity Neurosis: desire to recreate the world in the administrator's image. For other administrators' neuroses, see Chapters VI and VII.

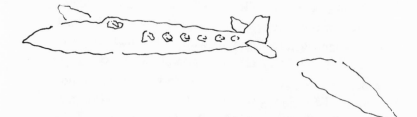

dangerous thing. Remember, even though a system may function very poorly, it still tends to Expand to Fill the Known Universe, and positive feedback only encourages that tendency.

Some radical humanists have suggested that people— ordinary human beings—should receive positive feedback (i.e., praise) for their *human qualities,* and that negative feedback to individuals should be avoided. In accordance with our basic Axioms, such a policy would cause people to tend to Expand to Fill the Known Universe, rather than Shrinking to Fit the System. The result, it may fairly be speculated, would be revolutionary, if not millenial. And since this is not a handbook for revolution, we cannot officially endorse such a policy. But we can suggest informally: try it, you might like it.

Oscillating Systems.

Alternating positive and negative feedback produces a special form of stability represented by endless oscillation between two polar states or conditions. In human systems, the phenomenon is perhaps best exemplified by certain committees whose recommendations slowly oscillate between two polar alternatives, usually over a period of some years. The period of the cycle has been found, on close examination, to be just longer than the time required for the advocates of Program "A" to graduate, retire, or otherwise leave the scene of action, being replaced by a new generation whose tendency is (having been exposed to the effects of Program "A") to demand Program "B." After some months or years of Program "B," a new generation of postulants is ripe to begin agitations to restore Program "A."

Thus, in academic affairs, Pass-Fail versus numerical or alphabetical grading represent two polar positions. In most universities, the Committee on Student Academic Affairs slowly oscillates from the one to the other of these two positions. The time period of this particular oscillation is just over four years, thus qualifying such Committees as Medium-Period Variables.

V. FUNCTIONARY'S FALSITY AND THE OPERATIONAL FALLACY

We move now to a consideration of an absolutely indispensable principle that must be understood by anyone who wishes to be considered adept in the field of Systemantics. It is not an easy theorem: it is subtle and elusive; and true mastery of it requires real intellectual effort. In order to make the road of the neophyte a little smoother as he struggles upward over this *pons asinorum* of systems theory, we approach it gradually and piecemeal, beginning with a lesser but closely related theorem.

Example. There is a man in our neighborhood who is building a boat in his backyard. He knows very little of boatbuilding and still less of sailing or navigation. He works from plans drawn up by himself. Nevertheless, he is demonstrably building a boat and can be called, in some real sense, a boatbuilder.

Now if you go down to Hampton Roads or any other shipyard and look around for a shipbuilder, you will be disappointed. You will find—in abundance—welders, carpenters, foremen, engineers, and many other specialists, but no shipbuilders. True, the company executives may call themselves shipbuilders, but if you observe them at their work, you will see that it really consists of writing

32

contracts, planning budgets, and other administrative activities. Clearly, they are not in any concrete sense building ships. In cold fact, a SYSTEM is building ships, and the SYSTEM is the shipbuilder. We conclude from the above that:

> People In Systems Do Not Do What The System
> Says They Are Doing.

This paradox was clearly recognized and described in detail by the nineteenth-century team of empirical sociologists, Gilbert and Sullivan, when they wrote:

> But that kind of ship so suited me
> That now I am the ruler of the Queen's Navee.

The "ship" referred to, the reader will recall, was the Admiral's legal *partner*ship.

Unfortunately, Gilbert and Sullivan, like so many pioneers of the prescientific era, failed to recognize the paradox as being of the essence of systems functions. By this oversight, the entire field was held back at least forty years.

In general, the larger and more complex the system, the less the resemblance between the true function and the name it bears. For brevity, we shall refer to this paradox as FUNCTIONARY'S FALSITY.* Further examples are legion. The following case studies may serve as warming-up exercises for the reader, who is urged to try out his own analytic skill:

*Advanced students are encouraged to use the term Korzybski's Semantic Anomaly.

1. What is the real-world function of a king?

Answer. In theory kings are supposed to rule their country, that is, to govern. In fact they spend much of their time and energy—and their country's treasure—fighting off usurpers. In democratic countries, a newly elected President may find it expedient to begin planning immediately for the next election.*

2. What is the real-world function of a university scholar?

Answer. University scholars are supposed to think and study deeply on basic intellectual problems of their own choosing. In fact, they must teach assigned courses, do "research" on problems for which research money is available, and publish, or perish.

By now the Systems-student should be sufficiently equipped to undertake, on his own, the analysis of such common Systems-functions as Professional Football Player (must include TV appearances), Minister of the Gospel (must be able to appear to advantage at White House breakfasts), or University President (fund-raising, riot-control).

The OPERATIONAL FALLACY is merely the Systems-analogue of FUNCTIONARY'S FALSITY. Just as PEOPLE IN SYSTEMS DO NOT DO WHAT THE SYSTEM SAYS THEY ARE DOING, so also THE SYSTEM ITSELF DOES NOT DO WHAT IT SAYS IT IS DOING. In slightly greater detail:

---

*On the other hand, a System designed to guarantee reelection may develop unexpected *bugs,* thus producing *creep*ing paralysis.

> The Function Performed By A System Is Not Operationally Identical To The Function Of The Same Name Performed By A Man. In General, A Function Performed By A Larger System Is Not Operationally Identical To The Function Of The Same Name As Performed By A Smaller System.

For example, you have a desire for a fresh apple.

(a) Nonsystems approach: You may (if you are very lucky and the season is right) stroll out of your farmhouse door and down to the orchard where you pick a dead-ripe, luscious specimen right off the tree.

(b) A *small system* serving the function with the same name (supplying a fresh apple) is the neighborhood grocery. Your grocer gets his apples in bushel baskets from the commercial orchard 20 miles away. The apples are not quite as fresh, and the very best of the lot have been selected for sale to gift houses, but you are still reasonably satisfied.

(c) A *large system* serving the "same" function is the supermarket chain. The apples are picked green and placed in "controlled atmosphere" storage where they tend to ripen, although the ripening process is now by no means the same as tree-ripening. The apples are then shipped by rail, spending days or weeks in boxcars under variable temperatures. The resulting product is called a "fresh apple" but in texture and flavor it bears little resemblance to (a) above.

The Function (Or Product) Is Defined By The Systems-Operations That Occur In Its Performance Or Manufacture.

The importance of Korzybski's contribution to understanding systems is now apparent. An apple that has been processed through the supermarket system is not the same as an apple picked dead ripe off the tree, and we are in error to use the same word for two different things.*

---

*We shall not attempt to pursue the origins of this sloppy semantic habit back to medieval scholasticism, which was more interested in the general essences of things than in their particularity. Nor shall we mention Plato, to whom only the essence was *really* real. Presumably, Plato had a plentiful supply of fresh apples in season and didn't have to worry about particulars.

A further point, of utmost importance for what comes after, is that *most of the things we human beings desire are nonsystems things.* We want a fresh apple picked dead ripe off the tree. But this is precisely what a large system can never supply. No one is going to set up a large system in order to supply one person with a fresh apple picked right off the tree. The system has other goals and other people in mind.

Apparent exceptions to the Operational Fallacy can be found in abundance. The true state of affairs, however, will not escape the discerning eye of the reader of this text who has taken his lesson to heart.

Example 1. Doesn't the auto industry supply us with millions of new cars each year, even tailoring them to our changing tastes in style and performance?

Answer. The reason we think the auto industry is meeting our needs is that we have almost completely forgotten what we originally wanted, namely, a means of going from one place to another that would be cheap, easy, convenient, safe, and fast. We have been brainwashed into thinking that the Detroit product meets these requirements.

If Detroit Makes It, It Must Be An Automobile*

Obviously, if what we desire is what a system is actually producing, the system will appear to be functioning as we wished.

Example 2. Doesn't the universal availability of cheap, fresh, enriched white bread represent a great systems

*A Systems-Delusion. See Chapter VI.

achievement in terms of nourishing the American population?

Answer. The short answer is that it is not bread. The French peasant eats fresher bread than we do, and it tastes better. The Egyptian fellah, one of the poorest farmers in the world, eats bread that is still hot from the oven at a price he can easily afford. Most of the cost of our bread is middleman costs—costs which would not be incurred if it were produced by local bakers rather than by a giant system.

The reader who has mastered Korzybski's Semantic Anomaly and the Operational Fallacy will be able to absorb without excitement, and even to nod knowingly at, the latest examples of its operation as reported in the daily press and on television. He will smile on being informed (as if it were something unexpected) that Leadership Training Fails to Train Leaders. He will quickly grasp the truth that the Boy Scout movement, designed to popularize camping out in the wilderness, has actually popularized—Scouting. And finally (that saddest example of the Operational Fallacy), he will understand the meaning of the fact that every peace-keeping scheme ever devised has included—as an essential component—an army.

VI. INSIDE SYSTEMS

Our study of the Operational Fallacy has made clear how and why it is that (1) large systems really do not do what they purport to do and that (2) people in systems are not actually performing the functions ascribed to them. These two facts quite naturally lend an air of unreality to the entire operation of a large system. In the present Chapter we propose to explore in more detail some of the effects of that unreal atmosphere, especially its effects upon the people in the system itself. But first, following our rule of placing the most fundamental Axioms at the beginning, we present the Fundamental Law of Administrative Workings (F.L.A.W.):

Things Are What They Are Reported To Be.

This Axiom has been stated in various ways, all to the same effect. Standard formulations include the following:

The Real World Is What Is Reported To The System

If It Isn't Official, It Hasn't Happened.

This proposition, which is fundamental to communication theory as well as to epistemology, may have been foreshadowed by Marshall McLuhan,* although he seems to have gotten it backwards: correctly stated, *The Message is the Medium* by which the System knows the World.

The observant Systems-student will no doubt be able to supply a number of variant readings of the same Law, gleaned from the newspapers and his own observations of government officials, corporation executives, et al. The net effect of this Law is to ensure that people in systems are never dealing with the real world that the rest of us have to live in but with a filtered, distorted, and censored version which is all that can get past the sensory organs of the system itself.

Corollary No. 1:

A System Is No Better Than Its Sensory Organs.

This Corollary, the validity of which is crystal clear to you and me, is viewed with perplexity by the personnel living within the system. For them, the real world *is* simply what their intake says it is, and any other world is only a wild hypothesis. A true Systems-person can no more imagine inadequacy of sensory function than a Flatlander can imagine three-dimensional space.

Corollary No. 2:

To Those Within A System, The Outside Reality Tends To Pale And Disappear.

This effect has been studied in some detail by a small group of dedicated General Systemanticists. In an effort to

---

*With McLuhan it is difficult to be sure about anything. The reader seeking greater clarity is referred to the murky brilliance of McLuhan's own prose.

introduce quantitative methodology into this important area of research, they have paid particular attention to the amount of information that reaches, or fails to reach, the attention of the relevant administrative officer. The crucial variable, they have found, is the fraction:

$$\frac{R_o}{R_s}$$

where
R_o equals   the amount of reality which fails to reach the relevant administrative officer

and
R_s equals   the total amount of reality presented to the system.

The fraction $\frac{R_o}{R_s}$ varies from zero (full awareness of outside reality) to unity (no reality getting through). It is known, of course, as the COEFFICIENT OF FICTION.

Positive Feedback (P.F.) obviously competes with Reality (R) for input into the System. The higher the P.F., the larger the quantity of Reality which fails to gain entrance to the System (R_o) and the higher the C.F. In systems employing P.F., values of C.F. in excess of 0.99 have been recorded.* Examples include evangelistic religious movements, certain authoritarian governmental systems, and the executive suites of most large corporations.

A high C.F. has particular implications for the relationship between the System and an Individual Person (repre-

---

*In theory the C.F. may attain 1.00, but in practice removing the last shred of reality from the sensory input becomes increasingly difficult.

sented by the lower-case letter i).* We state the relationship as follows.
Corollary No. 3:

The Bigger The System, The Narrower And More Specialized The Interface With Individuals.

In *very large* systems, the relationship is not with the
individual at all but with his social security number, his
driver's license, or some other paper phantom.

In systems of *medium size,* some residual awareness of
the individual may still persist. A hopeful indication was
recently observed by the author in a medium-sized hospital. Taped to the wall of the nurses' station, just above the
Vital Signs Remote Sensing Console that enables the
nurses to record whether the patient is breathing and even
to take his pulse without actually going down the hall to
see him was the following hand-lettered reminder:

The Chart Is Not The Patient.

Unfortunately this slogan, with its humanistic implications, turned out to be misleading. The nurses were neither
attending the patient nor making notations on the charts.
They were in the hospital auditorium taking a course in
Interdisciplinary Function.**

Government agencies, on the other hand, qualify as
components of truly *large systems.* Nowhere on the wall

*In mathematics, i represents an imaginary quantity.
**Interdisciplinary Function: The art of correlating one's own
professional activities more and more with those of other professionals while actually doing less and less.

of any such agency has the author seen a notation to the effect that

The Dossier Is Not The Person

—nor does he expect ever to see such a notation.

At this point we must offer our emendation of the work of Dr. Laurence J. Peter, who rightly pointed out that people within the system rise to their Level of Incompetence, with results everywhere visible. But it must be clear that the strange behavior of people in systems is not solely nor even primarily the result of mere incompetence, any more than it is mere "criminality" that makes men commit crimes. Would that the solution were so simple! A large role must be ascribed to the F.L.A.W., which isolates the Systems-person from the everyday world. As we all know, sensory deprivation tends to produce hallucinations. Similarly, immersion in a system tends to produce an altered mental state that results in various bizarre malfunctions, recognizable to us but not to the people in the system. We therefore pause for a Definition.

> FUNCTIONARY'S FAULT: A complex set of malfunctions induced in a Systems-person by the System itself, and primarily attributable to sensory deprivation.

Several subtypes of *Functionary's Fault* are known.* Only two will be singled out for special mention here. We leave to others the task of elaborating the complete

*Parkinson's recognition of Injelititis (see p. viii) stands as a landmark in the early history of Systems-pathology.

nosology of disorders of thought and behavior produced by systems.

1. *Functionary's Pride.*

This disorder was already ancient when it was definitively characterized by W. Shakespeare as "the insolence of office." A kind of mania of self-esteem induced by titles and the illusion of power, it is so well known as to need no further description. However, this treatise claims distinction as the first to attribute the syndrome, not to inherent vice nor to maladministration, but to the operation of the F.L.A.W. in the system itself upon the officeholder.

2. *Hireling's Hypnosis.*

A trance-like state, a suspension of normal mental activity, induced by membership within a system.

Example. A large private medical clinic and hospital installed a computerized billing system. One day the system printed out a bill for exactly $111.11 for every one of the more than 50,000 persons who had attended the clinic during the preceding year. For the next several days the switchboard was jammed with calls from irate recipients of the erroneous bills. Emergency calls could not get through. Nearly ten thousand former patients took their business elsewhere, for a total loss to the clinic of almost a million dollars.

The person operating the computer system that day, as well as the office clerks, the programmer, and twelve employes hired to stuff, seal, and stamp envelopes—all had observed the striking identity of numbers on the bills but had done nothing to stop the error. *The system had hypnotized them.*

Delusion Systems Versus Systems Delusions.

"I'm going to fly to New York this afternoon," you say. But what you really do, after driving an hour to get to the airport, is to strap yourself into a coffin-like tube of sheet metal and remain almost immobile, except for being passively shaken about, for a period of some hours. You are not flying in any real sense. At best the airplane could be said to be flying, though certainly not in the sense that birds fly.

Why do we not say that you are laboring under a delusion? The answer is: because we share your set of beliefs. We are in your delusion system, and we are both victims of a *systems delusion.*

Systems-delusions are the delusion systems that are almost universal in our modern world. Wherever a system is, there is also a systems-delusion, the inevitable result of the Operational Fallacy and the F.L.A.W. in systems.

Two Systems-delusions deserve particular mention because of their practical importance.

1. *Manager's Mirage.* *

The belief that some event (usually called an "outcome") was actually caused by the operation of the System. For example: The Public School System is believed by some to be responsible for the literary works of Faulkner, Hemingway, and Arthur Miller, since it *taught them to write.* Similarly, the NIH is credited with the major biomedical advances of the past generation, since it *funded the research.* We generalize:

> The System Takes The Credit (For What Would
> Probably Have Happened Anyway).

2. *Orwell's Inversion.*

The confusion of Input and Output.

Example. A giant program to Conquer Cancer is begun. At the end of five years, cancer has not been conquered, but one thousand research papers have been published. In addition, one million copies of a pamphlet entitled "You and the War Against Cancer" have been distributed. Those publications will absolutely be regarded as Output rather than Input.

Systems-People.

The preceding considerations have provided convincing evidence that the *System has its effects on the people within it.* It isolates them, feeds them a distorted and partial version of the outside world, and gives them the illusion of power and effectiveness. But systems and people are

*Equivalent technical terms: Output Assimilation; Kudos Capture.

related in another, subtler way. A selective process goes on whereby systems attract and keep those people whose attributes are such as to make them adapted to life in the system:

Systems Attract Systems-People.

While Systems-people share certain attributes in common, each specific system tends to attract people with specific sets of attributes. For example, people who are attracted to auto racing are likely to be people who enjoy driving fast, tinkering with high-powered cars, and beating other people in fierce competition. The System calls forth those attributes in its members and rewards the extreme degrees of them. But a word of warning is in order. *A priori* guesses as to what traits are fostered by a given system are likely to be wrong. Furthermore, those traits are not necessarily conducive to successful operation of the System itself, e.g., the qualities necessary for being elected president do not include the ability to run the country.

Systems attract not only Systems-people who have attributes for success *within* the system. They also attract individuals who possess specialized attributes adapted to

allow them to thrive at the expense of the system, i.e., persons who *parasitize* them. As the barnacle attaches to the whale, these persons attach themselves to systems, getting a free ride and a free lunch as long as the system survives.

Efforts to remove parasitic Systems-people by means of screening committees, review boards, and competency examinations merely generate new job categories for such people to occupy. Only the ancient Egyptians, with their deep insight into human organizations, had the courage to provide the radical remedy: a dual bureaucracy in which each job was represented twice, once by the honorary officeholder, once by the actual executive.

VII. ELEMENTARY
SYSTEMS-FUNCTIONS

"I never ruled Russia. Ten thousand clerks ruled Russia!"

Thus spoke Czar Alexander on his deathbed. Was he right? Of course he was! And at what a price he purchased that deep and depressing insight! *There* was a system designed, if any system ever was, to be a tool in the hands of one man, to permit that one man to carry into effect his slightest whim. And what happened? The system would not work. Worse yet, Alexander, with all his absolute authority, was unable to make it work.*

If Czar Alexander had been inclined to philosophical reflection, he would doubtless have meditated upon the functional vagaries of systems—their propensity for failing to work when most needed and for working overtime when one could really do without them. He might even have attained insight into the basic Axiom of Systems-function, from which all others are ultimately derived:

*Impotent Potentate Syndrome—a rather straightforward example. Additional examples: Mohammed commanding the mountain to come to him; King Canute desiring the sea to recede; President Nixon ordering the Watergate mess to disappear.

Big Systems Either Work On Their Own Or They Don't. If They Don't, You Can't Make Them.

Ignorance of this basic Axiom is responsible for the most widespread and futile of all administrative errors— pushing harder on the nonfunctioning system to make it work better (Administrator's Anxiety):

Pushing On The System Doesn't Help.

You might as well try to bring the elevator up to your floor by pulling on the indicator lever or pounding the call button. In fact, as we shall show in another place:

Even Trying To Be Helpful Is A Delicate And Dangerous Undertaking.

We do not deny that occasionally the parts of the nonfunctioning system may be so disposed that a good swift kick will cause them to fall into place so that the system can resume normal function. Ordinarily, however, such a maneuver merely produces one last spasmodic effort, after which the system subsides into total immobility.*

With this by way of introduction, let us proceed to a more detailed analysis of Systems Function, Malfunction, and Nonfunction. First off:

*Vending Machine Fallacy. Compare Vendetta: a feud between a customer and a recalcitrant vending machine.

A Simple System May Or May Not Work.

Simple systems that work are rare and precious additions to the armamentarium of human technology. They should be treasured. Unfortunately, they often possess attributes of *instability* requiring special skill in their operation. For example: the common fishing pole with line and hook; the butterfly net; skis; the safety razor; and (in Western hands) the boomerang. But simple systems possessing the required attribute of stability do exist: the footrule, the plumb bob, the button and buttonhole system, to name a few. Among simple systems involving human associations, the *family* occupies a special place.*

Although many of the world's frustrations are rooted in the malfunctions of complex systems, it is important to remember that *Some Complex Systems Actually Function.* This statement is *not* an Axiom. It is an observation of a natural phenomenon. The obverse of the *Primal Scenario,* it is not a truism, nor is there anything in modern philosophy that requires it to be true. We accept it here as a *given,* and offer humble thanks. The correct attitude of thankfulness leads naturally to the following *Rule of Thumb:*

If A System Is Working, Leave It Alone.
Don't Change Anything!

But how does it come about, step by step, that some complex systems actually function? This question, to

---

*See Chapter X: Practical Systems Design.

which we as students of General Systemantics attach the highest importance, has not yet yielded to intensive modern methods of investigation and analysis. As of this writing, only a partial and limited breakthrough can be reported, as follows:

A Complex System That Works Is Invariably Found To Have Evolved From A Simple System That Worked.

The parallel proposition also appears to be true:

A Complex System Designed From Scratch Never Works And Cannot Be Made To Work. You Have To Start Over, Beginning With A Working Simple System.

Diligent search for exceptions to these Axioms has yielded negative results.* The League of Nations? No. The United Nations? Hardly. Nevertheless, the conviction persists among some that a working complex system will be found somewhere to have been established *de novo,* from scratch. Mathematicians and engineers, in particular, insist that these formulations are too sweeping; that they set forth as natural law what is merely the result of certain technical difficulties, which they propose to overcome in the near future.**

Without committing ourselves too strongly to either

*Readers are invited to participate in the search and to report results (see Appendix III).
**Space for publication of their reports will be reserved in the *Journal of Negative Results.*

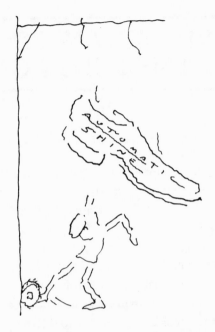

camp, we will remark that the mechanism by which the transition from *working simple system* to *working complex system* takes place is not known. Few areas offer greater potential reward for truly first-rate research.

VIII. ADVANCED
SYSTEMS-FUNCTIONS

In the preceding chapter we introduced certain elementary principles relating to the function or nonfunction of systems. We move now to more advanced concepts, some of which require close attention if they are to be mastered. The student will remember that our goal is twofold: first, to present the subject matter with rigorous logical correctness, moving sequentially from simple to more advanced ideas; and second, to provide a groundwork for *practical* mastery of the subject, so that the attentive student may deal with systems with the strength that comes from understanding. On the one hand, why flounder in unequal struggle when you can know in advance that your efforts will be unavailing? Nothing is more useless than struggling against a law of nature. On the other hand, there are circumstances (highly unusual and narrowly defined, of course) when one's knowledge of Systems-Functions will provide precisely the measure of extra added ability needed to tip the scales of a doubtful operation in one's favor. Those rare moments are, to the serious Systems-student, the reward and the payoff that makes worthwhile the entire long period of disciplined study and self-denial involved in mastery of this complex subject.

In accordance with our practice of moving from the simpler and more easily understandable to the more profound and impalpable, we present the following FUNCTIONAL INDETERMINACY THEOREM (F.I.T.):

In Complex Systems, Malfunction And Even
Total Nonfunction May Not Be Detectable For
Long Periods, If Ever.

This theorem often elicits surprise when first propounded. However, illustrative examples abound, especially from history. For example, it would seem reasonable to suppose that absolute monarchies, oriental despotisms, and other governments in which all power is concentrated in the will of one man would require as a minimum for the adequate functioning of those governments, that the will of the despot be intact. Nevertheless, the list of absolute monarchs who were hopelessly incompetent, even insane, is surprisingly long. They ruled with utter caprice, not to say whimsicality, for decades on end, and the net result to their countries was—undetectably different from the rule of the wisest kings.*

On the other hand, in strict accordance with the Generalized Uncertainty Principle, the greatest and wisest kings

---

*Students wishing to investigate this fascinating topic in more detail are advised to study the lives of Henry VIII, George III, certain Emperors of Japan, the Czars of Russia, the Sultans of Turkey, etc., etc. Readers may also wish to review the performances of present-day heads of state—the author wisely refrains.

have made decisions that proved disastrous. Charle-
magne, for example, in his desire to be fair to his three
sons, divided his empire among them—an act that gave
rise to France, Germany, and Alsace-Lorraine, and eventually to two World Wars.

The problem of evaluating "success" or "failure" as
applied to large systems is compounded by the difficulty
of finding proper criteria for such evaluation. What is the
system really supposed to be doing? For example: was the
Feudal System a "success" or a "failure"? Shall it be called
a "success" because it achieved the physical survival of
Western civilization, or shall it be called a "failure" because in place of the internationalism of Rome it be-
queathed us the doubtful legacy of nationalism and a divided Europe? Some thinkers, overwhelmed by the
difficulties of answering such questions, have taken refuge
in a Theorem of doubtful validity, which is presented here
for completeness' sake, without any commitment as to its
ultimate correctness:

Complex Systems Are Beyond Human Capacity To Evaluate

For our own part, we shall be content to quote the
words of one of the wisest of the Systems-Thinkers, as
follows:

> In general, we can say that the larger the system
> becomes, the more the parts interact, the more diffi-
> cult it is to understand environmental constraints,
> the more obscure becomes the problem of what re-

sources should be made available, and deepest of all, the more difficult becomes the problem of the legitimate values of the system.*

But, however difficult it may be to know what a system is doing, or even whether it is doing anything, we can still be sure of the validity of the Newtonian Law of Inertia as applied to systems:

A System That Performs A Certain Function Or Operates In A Certain Way Will Continue To Operate In That Way Regardless Of The Need Or Of Changed Conditions.

In accordance with this principle, the Selective Service System continued to register young men for the draft, long after the draft had ended.

When a System continues to do its own thing, regardless of circumstances, we may be sure that it is acting in pursuit of *inner goals.* This observation leads us, by a natural extension, to the insight that:

*C. West Churchman. *The Systems Approach.* New York: Dell Publishing Co., 1968, p. 77.

Systems Develop Goals Of Their Own The Instant
They Come Into Being.

Furthermore, it seems axiomatically clear that:

Intrasystem Goals Come First.

The reader who masters this powerful Axiom can read-
ily comprehend why the United Nations recently sus-
pended, for an entire day, its efforts at dealing with
drought, détente, and desert oil, in order to debate
whether UN employes should continue to ride first class
on airplanes.

We have used the terms *Goal, Purpose,* and *Function* a
number of times. In the interests of clarity, we here
sharpen our distinctions so as to recognize three separate
entities.

1. The Stated *Purpose* of the System (The "Goal" of the
Designer or Manager.)

2. The Built-in *Function* (What the System really does).

3. The *Goals* of the System Itself.

In an ideal world, there would be trivial differences only
among these three things, and the Science of General
Systemantics would be much simpler than it is. But we
must deal with things as they are, and at the moment we
are talking about Number 3, The Goals of the System
Itself.

Prior to and underlying any other Goal, the System has
a blind, instinctive urge to *maintain itself.* In Axiomatic
form:

The System Has A Will To Live.

We have stated this Axiom in subjective terms redolent of teleology, to which some philosophers may object. As we seek no tendentious arguments with them or with anyone else, we willingly retreat to a more neutral formulation and assert simply:

The System Behaves As If It Has A Will To Live.

This natural tendency on the part of Systems, coupled with the F.L.A.W. and the power of Systems-Delusions, often causes managers, political leaders, and other Systems-persons to produce statements of the general form:
"What's good for General Motors is good for the Country."

Such propositions merely reflect the Systems-bias of their authors. They may be momentarily accurate, but they cannot be *generally* correct in any logical or scientific sense. Only one proposition of this form is *necessarily* true. With modest pride we state it as follows:

What's Good For General Systemantics Is Good
For The Country.

IX. SYSTEMS-FAILURE
(THEORY OF ERRORS)

In the early days of the development of electronic computers, engineers were startled to observe that the probability of malfunction was proportional to the *fourth power* of the number of vacuum tubes. That observation led them to a preoccupation with component reliability that has culminated in modern transistorized, solid-state computers. But, impressive as that development may be, it is, from the standpoint of general Systemantics, merely a neurotic digression from the straight and narrow path. Improvement in component reliability merely postpones the day of reckoning. As the system grows in size and complexity, it gradually but inevitably outgrows its component specifications. *Parts* (whether human or electronic) *begin to fail.* The important point is:

> Any Large System Is Going To Be Operating
> Most Of The Time In Failure Mode.

What the system is supposed to be doing when everything is working well is really beside the point because that happy state is never achieved in real life. The truly pertinent question is: How does it work when its components aren't working well? *How does it fail? How well does it function in failure mode?*

61

Our basic approach is indicated in the Fundamental Failure Theorem (F.F.T.):

A System Can Fail In An Infinite Number Of Ways.

An extreme example is the government of Haiti, which, with one exception, appears to consist entirely of departments that do not function.* Dozens of national and international aid agencies, frustrated by the inability of the Haitian government to cope with outside assistance, have sent emergency representatives to Haiti, to teach the government officials *how to fill out requests for aid.*

Purists and theoreticians may argue that the number of ways in which a system can fail is not truly *infinite,* but merely *very large.* While conceding the theoretical interest of such speculations, we hold ourselves aloof from the dust of polemical strife. More to the point of our investigations, which are oriented toward the world of practicalities, is the fact that, while some kinds of failure may be easily predictable, most are not. In general:

The Mode Of Failure Of A Complex System Cannot Ordinarily Be Predicted From Its Structure.

Strictly speaking, this proposition is merely a corollary of the GENERALIZED UNCERTAINTY PRINCIPLE; but, because of its practical importance, it is here elevated to the status of an Axiom. Beginners in science

---

*The tobacco tax, which goes directly into the President's personal bank account, is meticulously collected in full.

and in politics commonly deny the truth of this Axiom until its validity has been repeatedly borne in upon them by repeated disasters attendant upon their own pet schemes. Some, indeed, in both professions, persist in their Systems-delusional views to the very end of life. As a result, it is usually the case that:

The Crucial Variables Are Discovered By Accident.

Example 1. The Pyramid of Snofru. On the edge of the desert, a few miles south of the Great Pyramids of Egypt, stands a ruined tower of masonry some two hundred feet high, surrounded by great mounds of rubble. It is the remains of a gigantic pyramid. Its ruined state has been variously attributed to time, weather, earthquake, or vandalism despite the obvious fact that none of these factors has been able to affect the other Great Pyramids to anywhere near the same degree.

Only in our own time has the correct solution to this enigma been advanced. In conformity with basic Systems-principles* the answer was provided by an outsider, a physicist unaware that there was any problem, who, after a vacation in Egypt, realized that the pyramid of Snofru had *fallen down.* The immense piles of rubble are big enough to reconstruct the entire pyramid. It is clear that the thing was almost complete when it fell.

Why did Snofru's pyramid fall down when that of his grandfather Zoser, which had the form of a stepped tower, did not? The answer (provided by our visiting physicist)

*See "The 'Problem' Problem," Chapter XI.

is that Snofru tried to face his tower with stone blocks to create a true pyramid with smoothly sloping sides. Unknown to Snofru, the crucial factor was just this: a stepped tower of those gigantic dimensions is stable, a true pyramid is not. *It fell down.* *

Example 2. The Pyramid of Cheops. Cheops, son of Snofru, vowed not to make the same mistake. With great care he constructed his pyramid of finely dressed limestone blocks, mathematically placed to distribute the stresses. His pyramid did not fall down, nor did those of his immediate successors, which were built in the same way. But the Egyptian state, subjected to unbearable stresses by the building of those monsters of pride, collapsed into anarchy. *Egypt fell down.*

The Fail-Safe Theorem.

In expounding the science of General Systemantics we have striven to remain at the level of fundamental theory, providing the reader with basic insights of broad general validity, and stopping by the wayside only occasionally to pluck a Paradox or collar a Corollary of special interest for specific situations. We have not tried to tell the reader how to run his own affairs or those of the particular community, society, or nation to which he may happen to belong. We will, in general, continue to adhere to that policy. But at this point in our exposition we deliberately swerve from our splendid isolation to inject a note of exhortation, even of urgency, into the discussion. We believe the Fail-Safe Theorem to be of special immediacy for everyone concerned with the fate of Homo sapiens, and we

---

*Mendelssohn, Kurt. "A Scientist Looks at the Pyramids." *American Scientist,* 59: 210–220 (March-April), 1971.

therefore urge all readers to read, mark, learn, and inwardly digest it, to the end that future policy may not ignore it but rather take it and its implications into account. Tedious exegesis would be insulting to readers of this work and, worse, boring. We give it, therefore, in all its austere and forbidding simplicity:

> When A Fail-Safe System Fails, It Fails By
> Failing To Fail Safe.

Nuclear strategists please note.

X. APPLIED SYSTEMANTICS 1: PRACTICAL SYSTEMS DESIGN

Up to this point our mode of development of the subject has been rigorously logical, austerely theoretical. We have made few if any concessions to mere expediency or the demands of the market place. The earnest Systems-student who has toiled thus far may justifiably ask: Is this all, this handful of abstractions? Did not the author promise us, way back in the Preface, that we would gain not merely in understanding but also in practical know-how if we but applied ourselves to mastery of this difficult science?

True, the insights of practical wisdom so derived are pitifully few, and their power is strictly limited. You have already mastered the hard lesson of the Operational Fallacy, with its dreary implication that SYSTEMS NEVER REALLY DO WHAT WE WANT THEM TO DO. What is now offered is a primer of What Can Be Done, given the existence of such built-in limitations. Briefly, what can be done is to *cope,* and, on rare and satisfying occasions, to *prevail.* The mature and realistic Systems-student asks no more than this of life, taking his contentment primarily from the fact that he has lived in accord

67

with Nature herself, and only secondarily from the fact that no one else can get any more, either.

It is hardly necessary to state that the very first principle of Systems-design is a negative one:

Do It Without A System If You Can.

For those who need reasons for such a self-evident proposition, we offer the following concise summary of the entire field of General Systemantics.

Systems are seductive. They promise to do a hard job faster, better, and more easily than you could do it by yourself. But if you set up a system, you are likely to find your time and effort now being consumed in the care and feeding of the system itself.

New problems are created by its very presence.*

Once set up, it won't go away, it grows and encroaches.**

It begins to do strange and wonderful things.***

Breaks down in ways you never thought possible.****

It kicks back, gets in the way, and opposes its own proper function.*****

Your own perspective becomes distorted by being in the system.******

You become anxious and push on it to make it work.*******

*Fundamental Theorem.
**Laws of Growth.
***Generalized Uncertainty Principle (G.U.P.)
****Fundamental Failure Theorem (F.F.T.)
*****Le Chatelier's Principle.
******Functionary's Fault.
*******Administrator's Anxiety.

Eventually you come to believe that the misbegotten product it so grudgingly delivers is what you really wanted all the time.*

At that point encroachment has become complete. You have become absorbed. You are now a Systems-person.

Furthermore, some things just can't be done well by a system. In formal language:

Many Functions Are Intrinsically Unsuited To The Systems Approach.

Anyone who has tried to manipulate an umbrella in a high wind has an intuitive feel for what is involved here. Technically, it has something to do with rapid and irregular fluctuations in the parameters.** The great secret of Systems Design is to be able to sense what things can naturally be done easily and elegantly by means of a system and what things are hard—and to stay away from the hard things.

Also, if you really *must* build a system, by all means remember the Systems Law of Gravity, otherwise known as the Vector Theory of Systems:

---

*Systems-delusion.
**Parameters are variables travelling under an assumed name. They are variables that are held mathematically constant, which just goes to show how little mathematics knows about the real world.

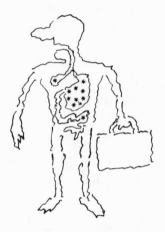

Systems Run Best When Designed To Run Downhill.

In human terms, this means working *with* human tendencies rather than *against* them. For example, a state-run lottery flourishes even in times of economic depression because its function is aligned with the basic human instinct to gamble a small stake in hopes of a large reward. The public school system, on the other hand, although founded with the highest and most altruistic goals in mind, remains in a state of chronic failure because it violates the human principle of spontaneity. It goes against the grain, and therefore it does not ever really succeed.

Finally, don't make the system too tight. This is usually done in the name of efficiency, or (paradoxically) in the hope of making the system more permanent. Neither goal is achieved if the resulting system (a) doesn't work at all; (b) disintegrates; or (c) rapidly loses steam and peters out:

Loose Systems Last Longer And Function Better.

Consider, for example, the System of the Family. The family has been around for a long time. Our close primate relatives, the gorillas, form family units consisting of husband and wife and one or more offspring. As Jane Goodall has shown, gorillas take naps after meals. (Every day is Sunday for large primates.) The youngsters wake up too soon, get bored and start monkeying around the nest. Father gorilla eventually wakes up, leans on one elbow, and fixes the errant youngster with a penetrating stare that speaks louder than words. The offending juvenile thereupon stops his irritating hyperactivity, at least for a few minutes.

Clearly, this is a functioning family system. Its immense survival power is obvious. It has endured vicissitudes compared to which the stresses our own day puts on it are trivial. And what are the sources of its strength? They are extreme simplicity of structure; looseness in everyday functioning; "inefficiency" in the efficiency expert's sense of the term; and a strong alignment with basic primate motivations.*

---

*Even in its remotest origins, however, the Family System exacts its price. What Father Gorilla wanted is not exactly what he gets. There is a tradeoff for the continued presence of the female—namely, the continued presence of the offspring. The Operational Fallacy is not to be denied.

XI. APPLIED SYSTEMANTICS 2: MANAGEMENT AND OTHER MYTHS

Our previous encounter with Czar Alexander and the Impotent Potentate Syndrome should have alerted us already to the pitfalls of Systems-management. Should we actually find ourselves, against our better judgment, in a managerial or executive position, we should remember the dread effects of the F.L.A.W., of Hireling's Hypnosis, and of the pervasive Systems-delusions. The combined effect of those forces is such as to render very doubtful any form of Management Science. If there is no way of being sure what the real effect of your managerial actions has been, how can you know whether you have done well or ill?

The F.L.A.W. may even operate in such a way as to hide from the administrator the operation of the G.U.P. In such situations, the administrator sinks into complacency while his System careens from disaster to disaster. In recognition of major Russian contributions to our experience of this phenomenon, it is known as the Potemkin Village Effect. The P.V.E. is especially pronounced in Five-Year Plans, which typically report sensational over-achievement during the first four and a half years, followed by a rash of criminal trials of the top officials and the announcement of a new and more ambitious Five-Year Plan, starting from a baseline somewhat *lower* than that of the preceding Plan, but with *higher goals.*

Catalytic Managership.

Although in a theoretical work of this type we cannot stop to offer specific advice, we can suggest that the Systems-sophisticated Manager will adopt a style which we designate as Catalytic Managership. Briefly, Catalytic Managership is based on the premise that trying to make something happen is too ambitious and usually fails, resulting in a great deal of wasted effort and lowered morale. It is, however, sometimes possible to remove obstacles in the way of something happening. A great deal may then occur with little effort on the part of the manager, who nevertheless gets a large part of the credit. The situation is similar to that of the lumberjack who picks out the key log from a logjam, or the chemist who adds the final pinch of reagent to an unstable mixture. But a warning is in order. Catalytic Managership will only work if the System is so designed that Something Can Actually Happen—a condition that commonly is not met.

Catalytic Managership has been practiced by leaders of genius throughout recorded history. M. Gandhi is reported to have said: "There go my people, I must run and catch up with them in order to lead them." Choosing the correct System is crucial for success in Catalytic Managership—consider the probable career of W. Churchill had he been Prime Minister of Angola.

The "Problem" Problem.

For the practicing Systems-manager, the greatest pitfall lies in the realm of Problems and Problem-solving. Systems can do many things, but one thing they emphatically cannot do is to Solve Problems. This is because Problem-solving is not a Systems-function, and there is no satisfactory Systems-approximation to the solution of a Problem. A System represents someone's solution to a problem. The System does not *solve* the problem. Yet, whenever a particular problem is large enough and puzzling enough to be considered a Capital–P Problem, men rush in to solve it by means of a System.

Once a problem is recognized as a Problem, it undergoes a subtle metamorphosis. Experts in the "Problem" area proceed to elaborate its complexity. They design systems to attack it. *This approach guarantees failure,* at least for all but the most pedestrian tasks. A system that is sufficiently large, complex, and ambitious can reduce output far below "random" levels. Thus, a federal Program to conquer cancer may tie up all the competent researchers in the field, leaving the problem to be solved by someone else, typically a graduate student from the University of Tasmania doing a little recreational entomology on his vacation. *Solutions usually come from people who see in the problem only an interesting puzzle, and whose qualifications would never satisfy a select committee.*

Item. When Pasteur accepted the challenge of the French silk producers to discover the cause of silkworm disease, he had never seen, much less studied, a silkworm. He was not even a biologist.

Item. The Wright brothers, who built the first successful heavier-than-air machine, were *bicycle makers.*

Item. The molecular structure of the gene—closest thing to the true "secret of life"—was revealed through X-ray crystallography, a technique having little to do with biology. And James Watson, who solved the puzzle, was not an X-ray crystallographer. He was not even a chemist. Furthermore, he had been refused a renewal of his research grant because his sponsors felt he wasn't sticking to the point.

As these examples make clear, great advances may be achieved by individuals working essentially alone or in small teams. But what of the reverse situation? What is the track record of large systems designed for the express purpose of solving a major problem? A decent respect for our predecessors prevents us from dwelling upon the efforts of successive governmental administrations to eradicate poverty, reduce crime, or even get the mail delivered on time. The bankruptcy of the railroad system, already achieved under private management, has been further compounded with government assistance. In the field of science, a famed research center recently screened over fifty thousand different chemical substances, at great expense, for anticancer activity—with negative results. We conclude that:

> Great Advances Do Not Come Out Of Systems
> Designed To Produce Great Advances.

Furthermore:

> Complicated Systems Produce Complicated
> Responses (Not Solutions) To Problems.

Even more disastrous than the "Problem" approach to problems is the "Crash" approach. The "Crash" approach combines the adverse dynamics of the "Problem" approach with elements of Administrator's Anxiety (Pushing on the System to Make It Work) and plain hysteria. Under the pressures of such a System, scientists themselves (normally the most tranquil and reflective of men) may begin to crack, cutting off tails and painting the skins of mice in desperate efforts to meet the artificial but pressing goals of the System. The prevention of such disasters clearly calls for Catalytic Managership of the highest order.

XII. APPLIED SYSTEMANTICS 3: TAMING SYSTEMS

No treatise on Systems would be complete without some mention of those professional Systems-people who call themselves "change agents." The belief that constructive change in a system is possible through direct intervention is an optimistic view that dies hard. The alert Systems-student will recognize it as the common psychological phenomenon of *denial* of unpleasant reality. In fact, it may be viewed as a classic example of a Systems-delusion. Even more insidious, however, is the implicit assumption that there is a science of Systems-intervention which any diligent pupil can master, thereby achieving competence to intervene here and there in systems large and small with results objectively verifiable and of consistent benefit to mankind. Such a belief is clearly in conflict with the Generalized Uncertainty Principle.*

We do not take an absolutely pessimistic stand. It is possible, we believe, to exert some limited influence on large systems. But we resolutely assert that any such influ-

---

*Most change agents survive between jobs by writing magazine articles explaining the reasons for the disaster that struck the latest object of their change agentry.

ence must occur within the framework of, and in accordance with, the general laws of Systems-function enunciated in this treatise.

The work of change agents is made enormously more delicate and uncertain by the fact that the mere presence of a change agent (recognizable as such) has about the same effect on an organization as an efficiency expert seen strolling across the factory floor with stopwatch in hand: it promptly induces bizarre and unpredictable alterations in the behavior of the system. Because of this effect anyone who identifies himself publicly as a change agent automatically convicts himself of incompetence. Changes will certainly occur as a result, but they are not likely to be the changes desired.

Despite the built-in difficulties of change agentry, there are a few examples on record of situations in which a recalcitrant system has been tamed, i.e., the worst features of its behavior have been tempered so as to produce a tolerable result. How such interventions have come about is not at all clear. Most are shrouded in the obscurity of the distant past. What is clear is that the remedy must strike deeply at the roots of the system itself in order to produce any significant effect. Furthermore, an uncanny element of paradox is prominent in the few examples so far reported. Thus, the long survival of the British monarchy is probably attributable to the fact that the King reigns but does not rule. The cohesion of the far-flung dominions of the Empire is similarly based on the paradoxical fact of voluntary association.

An even more challenging example is the *Token System,* with which mankind has been having trouble ever since the Phoenicians invented it. K. Marx was perhaps the first to point out its defects as a system, thereby qualifying himself as a pioneer Systems-thinker.

Briefly, the Token System is intended to provide for distribution of wealth along certain rational lines, such as the contribution of the individual to the common welfare. In practice, however, and in accordance with an ineluctable natural law of Systems-behavior, the tokens are accumulated by those whose primary virtue is skill in accumulating tokens—a point overlooked by Marx.

A recital of the schemes devised by mankind to correct, or at least neutralize, this intrinsic difficulty with the Token System makes tedious and depressing reading indeed. There are some who go so far as to assert that modern history is mainly the story of those efforts. Governments everywhere, whether capitalist, socialist, or communist, have struggled to Tame the Token System. Only one society, hidden in a far-off corner of the world, has had the imagination and daring to achieve success in this effort. For the sake of our industrial civilization, sunk in the miseries of operation of the Token System, we here present our findings.

On the Island of Yap in the South Pacific, money is in the form of stone coins, as big as cartwheels, with a hole in the center. The value of a coin is based, not on size, but on the number of people who died bringing it across the open sea, lashed to the bow of a frail canoe, from the limestone quarries of the island of Palau, 250 miles from Yap.

No Yapese person can reasonably hope to accumulate any large number of such coins. In fact, when possession of a coin changes, the position of the coin itself does not. It continues to lie wherever it has always lain, along a path

This coin is worth twenty dead people. How many is your money worth?

or on the edge of a village. Only the abstract title changes, and nothing of consequence has changed for the Yapese people.

Clearly, there is no problem of theft or of hoarding.

The assignment of value on the basis of men lost on the journey is an additional stroke of genius. The coin cannot be used as a symbolic substitute for human labor. It does not represent so many coconuts collected, so many pounds of copra produced, or so many head of cattle or chickens. No one can, by accumulating tokens, hold the community to ransom.

Critics may argue that this cure of the faults of the Token System is too radical—that by depriving coinage of the two attributes of portability and symbolic representation of human labor, the Yapese have in fact "de-monetized" their currency, so that it is no longer money. Against this hyperfine argument we place the simple observation that everyone, everywhere immediately recognizes the Yapese coins for what they are—real money. It will take more than the quibbles of specialists to convince the average man that what he sees with his own eyes is not the truth.

Can the Government System be Tamed?

Students of General Systemantics will have apprehended by now that General Systemantics does not offer ready-made formulas for the solution of Systems-problems, even of such pressing problems as Warfare between Nations or Governmental Oppression. The Axioms are too fundamental for direct application to practical situations, and the intervening methodology has in any case not been worked out. At most, one may derive a clue to a method of approach whereby the *Intrinsic Difficulty* is specified as precisely as possible, so that daring and imaginative correctives may be tried. The risk of failure or even of catastrophe is very high, and the undertaking should be begun only where the present evil is very clear and the consequences of a miscarriage are judged to be no more unbearable than a continuation of the original unsatisfactory situation.

With these reservations, we may permit ourselves a bit of harmless speculation on the Government System. Government Systems, acting in accordance with the Laws of Growth, Tend to Expand and Encroach. In encroaching upon their own citizens, they produce Tyranny, and in encroaching on other Government Systems, they engage in Warfare. If one could correctly identify the Intrinsic Difficulty with the Government System, one might be able to curb or neutralize those two tendencies, to the benefit of the Mankind System.

What is the Intrinsic Difficulty with the Government System? Previous reformers, identifying the core problem as the *concentration of power* in a few hands, have attempted to improve things by diffusing that power. This works temporarily, but gradually (Systems Law of Gravity) the power becomes concentrated again.

A breakaway group of General Systemanticists, starting from the principle that it is very difficult to unscramble eggs, have proposed that the core problem is not the concentration of power but the concentration of the governed in one place, where the government can get at them. They have proposed, not the diffusion of power, but the diffusion of the targets of power—the citizens themselves.

They would achieve this by providing citizens with *two*

new freedoms, in addition to the traditional Four Freedoms. These two new freedoms, appropriately designated as the Fifth and Sixth Freedoms, are:

(5) Free Choice of Territory (Distributional Freedom).

(6) Free Choice of Government (Principle of Hegemonic Indeterminacy).

Under Free Choice of Territory, a citizen of any country is free to live in any part of the world he chooses. He remains a citizen of the government he prefers, to which he pays taxes and for whose officers he votes. However, as the term Free Choice of Government implies, he may at any time change his citizenship and his allegiance from his present government to another government that offers more attractive tax rates, better pensions, more interesting public officials, or simply an invigorating change of pace.*

With these two new Freedoms in effect, one would expect that after a short period of equilibration, citizens of any nation would be distributed amongst the citizens of all other nations—not necessarily at random, but sufficiently so for our purpose, which is to remove them effectively from the grip of their own government. A government can hardly put any large number of its own citizens in jail if it has to send halfway around the world for them, one by one, and persuade other governments of the justice of the proceedings. Raising armies would become administratively impossible. Furthermore, wars of one government against another would become impractical, since large numbers of the "enemy" would be distributed all over the

---

*Common courtesy would seem to require two weeks' advance notice; the standard notice any employer would give an employee.

world, including the territory of the home government.

The net result of the two new Freedoms would be to break up the Concentration of the Governed, to divide and distribute them throughout other governments, a principle which we shall call the *Comminution of Hegemony.* If practiced on a world-wide scale it could lead to revolutionary changes in the relationship of citizens to their governments, reversing the traditional polarity and making governments fearfully dependent upon the favor or even the whims of their citizenry rather than vice versa. In keeping with the revolutionary aspects of this proposal, we hereby broach the solemn question:

World Comminution: Threat Or Promise?

ENVOI:
BEYOND EXPERTISE

We have come to the end of our presentation; why not simply stop? Does Euclid bother to round off his *Elements* with a polished little essay on the significance of the whole work?* But, lest readers feel that they have been left hanging in air, so to speak, this coda is appended. We shall not review the purposes set forth in the Preface that motivated us to undertake this work, nor shall we describe at length how the intervening chapters have neatly covered all aspects of the topic. Instead, we shall speak to the necessity of a New Breed of Systems-student—one who, having absorbed the Axioms here collected and, more importantly, the *spirit* infusing them, can progress beyond technology to the kind of wisdom the world needs. The world already suffers from too many experts. They tell us more than we need to know or dare to ask about heavier-than-air machines, fusion bombs, and management science. What we really need to know is much more subtle.

The questions we must answer for good or ill are of a different order: Can placing a microphone in the Oval Office bring down the government? Will setting up Management by Objectives in the Universities bring on another Dark Age? Will permitting men and women everywhere

*Of course not.

85

the freedom to choose their own way of life and to make their own decisions lead to a better world? For such questions your run-of-the-mill expert is of no value. What is required is a special, elusive talent, really an intuition—a feel for the wild, weird, wonderful, and paradoxical ways of large systems. We offer no formula for recognizing or cultivating such a talent. But we suggest that its possessors will, more likely than not, have cut their eyeteeth on the Axioms of General Systemantics.

APPENDIX I:
ANNOTATED COMPENDIUM OF BASIC SYSTEMS-AXIOMS, THEOREMS, COROLLARIES, ETC.

For convenience and ready reference of both scholar and casual reader, we here summarize the results of our researches on General Systemantics in one condensed, synoptic tabulation. This is by no means to be considered as a comprehensive or all-inclusive listing of all possible Systems-propositions. Only the most basic are included. Now that the trail has been blazed, readers will surely find numerous additional formulations in every field of science, industry, and human affairs. An undertaking of special interest to some will be that of translating certain well-known laws in various fields into their completely general Systems-formulations. You, the reader, are invited to share with us your own findings (see Readers' Tear-Out Feedback Sheet, Appendix III).

Annotated Compendium

1. The Primal Scenario or Basic Datum of Experience:
   SYSTEMS IN GENERAL WORK POORLY OR NOT AT ALL.
   Alternative formulations:
   NOTHING COMPLICATED WORKS.
   COMPLICATED SYSTEMS SELDOM EXCEED 5 PERCENT EFFICIENCY.
   (There is always a fly in the ointment.)*

---

*Expressions in parentheses represent popular sayings, proverbs, or vulgar forms of the Axioms.

2. The Fundamental Theorem:
   NEW SYSTEMS GENERATE NEW PROBLEMS.
   Corollary (Occam's Razor):
   SYSTEMS SHOULD NOT BE UNNECESSARILY
   MULTIPLIED.

3. The Law of Conservation of Anergy:
   THE TOTAL AMOUNT OF ANERGY IN THE
   UNIVERSE IS CONSTANT.
   Corollary:
   SYSTEMS OPERATE BY REDISTRIBUTING AN-
   ERGY INTO DIFFERENT FORMS AND INTO
   ACCUMULATIONS OF DIFFERENT SIZES.

4. Laws of Growth:
   SYSTEMS TEND TO GROW, AND AS THEY
   GROW, THEY ENCROACH.
   Alternative Form—The Big-Bang Theorem of Systems-
      Cosmology:
   SYSTEMS TEND TO EXPAND TO FILL THE
   KNOWN UNIVERSE.
   A more conservative formulation of the same principle is
      known as Parkinson's Extended Law:
   THE SYSTEM ITSELF TENDS TO EXPAND AT
   5–6 PERCENT PER ANNUM.

5. The Generalized Uncertainty Principle:
   SYSTEMS DISPLAY ANTICS.
   Alternative formulations:
   COMPLICATED SYSTEMS PRODUCE UNEX-
   PECTED OUTCOMES.
   THE TOTAL BEHAVIOR OF LARGE SYSTEMS
   CANNOT BE PREDICTED.
   (In Systems work, you never know where you are.)
   Corollary: The Non-Additivity Theorem of Systems-
      Behavior (Climax Design Theorem):

A LARGE SYSTEM, PRODUCED BY EXPAND-
ING THE DIMENSIONS OF A SMALLER SYS-
TEM, DOES NOT BEHAVE LIKE THE SMALLER
SYSTEM.

6. Le Chatelier's Principle:
COMPLEX SYSTEMS TEND TO OPPOSE THEIR
OWN PROPER FUNCTION.
Alternative formulations:
SYSTEMS GET IN THE WAY.
THE SYSTEM ALWAYS KICKS BACK.
Corollary:
POSITIVE FEEDBACK IS DANGEROUS.

7. Functionary's Falsity:
PEOPLE IN SYSTEMS DO NOT DO WHAT THE
SYSTEM SAYS THEY ARE DOING.

8. The Operational Fallacy:
THE SYSTEM ITSELF DOES NOT DO WHAT IT
SAYS IT IS DOING.
Long Form of the Operational Fallacy:
THE FUNCTION PERFORMED BY A SYSTEM IS
NOT OPERATIONALLY IDENTICAL TO THE
FUNCTION OF THE SAME NAME PERFORMED
BY A MAN. *In general:* A FUNCTION PER-
FORMED BY A LARGER SYSTEM IS NOT OPER-
ATIONALLY IDENTICAL TO THE FUNCTION
OF THE SAME NAME PERFORMED BY A
SMALLER SYSTEM.

9. The Fundamental Law of Administrative Workings
(F.L.A.W.):
THINGS ARE WHAT THEY ARE REPORTED TO
BE.
THE REAL WORLD IS WHATEVER IS RE-
PORTED TO THE SYSTEM.

(If it isn't official, it didn't happen.)
Systems-delusion:
(If it's made in Detroit, it must be an automobile.)
Corollaries:
A SYSTEM IS NO BETTER THAN ITS SENSORY
ORGANS.
TO THOSE WITHIN A SYSTEM, THE OUTSIDE
REALITY TENDS TO PALE AND DISAPPEAR.

10. SYSTEMS ATTRACT SYSTEMS-PEOPLE.
Corollary:
FOR EVERY HUMAN SYSTEM, THERE IS A
TYPE OF PERSON ADAPTED TO THRIVE ON IT
OR IN IT.

11. THE BIGGER THE SYSTEM, THE NARROWER
AND MORE SPECIALIZED THE INTERFACE
WITH INDIVIDUALS.

12. A COMPLEX SYSTEM CANNOT BE "MADE"
TO WORK. IT EITHER WORKS OR IT DOESN'T.
Corollary (Administrator's Anxiety):
PUSHING ON THE SYSTEM DOESN'T HELP. IT
JUST MAKES THINGS WORSE.

13. A SIMPLE SYSTEM, DESIGNED FROM
SCRATCH, SOMETIMES WORKS.
Alternatively:
A SIMPLE SYSTEM MAY OR MAY NOT WORK.

14. SOME COMPLEX SYSTEMS ACTUALLY
WORK.
Rule Of Thumb:
IF A SYSTEM IS WORKING, LEAVE IT ALONE.

15. A COMPLEX SYSTEM THAT WORKS IS IN-
VARIABLY FOUND TO HAVE EVOLVED FROM
A SIMPLE SYSTEM THAT WORKS.

16. A COMPLEX SYSTEM DESIGNED FROM SCRATCH NEVER WORKS AND CANNOT BE PATCHED UP TO MAKE IT WORK. YOU HAVE TO START OVER, BEGINNING WITH A WORKING SIMPLE SYSTEM.
Translation for Computer-Programmers:
PROGRAMS NEVER RUN THE FIRST TIME. COMPLEX PROGRAMS NEVER RUN.
(Anything worth doing once will probably have to be done twice.)

17. The Functional Indeterminacy Theorem (F.I.T.):
IN COMPLEX SYSTEMS, MALFUNCTION AND EVEN TOTAL NONFUNCTION MAY NOT BE DETECTABLE FOR LONG PERIODS, IF EVER.*

18. The Kantian Hypothesis (Know-Nothing Theorem):
LARGE COMPLEX SYSTEMS ARE BEYOND HUMAN CAPACITY TO EVALUATE.

19. The Newtonian Law of Systems-Inertia:
A SYSTEM THAT PERFORMS A CERTAIN FUNCTION OR OPERATES IN A CERTAIN WAY WILL CONTINUE TO OPERATE IN THAT WAY REGARDLESS OF THE NEED OR OF CHANGED CONDITIONS.
Alternatively:
A SYSTEM CONTINUES TO DO ITS THING, REGARDLESS OF NEED.

20. SYSTEMS DEVELOP GOALS OF THEIR OWN THE INSTANT THEY COME INTO BEING.

---

*Such systems may, however, persist indefinitely or even expand (see Laws of Growth, above).

21. INTRASYSTEM GOALS COME FIRST.

(The following seven theorems are referred to collectively as the Failure-Mode Theorems.)
22. The Fundamental Failure-Mode Theorem (F.F.T.): COMPLEX SYSTEMS USUALLY OPERATE IN FAILURE MODE.

23. A COMPLEX SYSTEM CAN FAIL IN AN INFINITE NUMBER OF WAYS.
(If anything can go wrong, it will.)

24. THE MODE OF FAILURE OF A COMPLEX SYSTEM CANNOT ORDINARILY BE PREDICTED FROM ITS STRUCTURE.

25. THE CRUCIAL VARIABLES ARE DISCOVERED BY ACCIDENT.

26. THE LARGER THE SYSTEM, THE GREATER THE PROBABILITY OF UNEXPECTED FAILURE.

27. "SUCCESS" OR "FUNCTION" IN ANY SYSTEM MAY BE FAILURE IN THE LARGER OR SMALLER SYSTEMS TO WHICH THE SYSTEM IS CONNECTED.
Corollary:
IN SETTING UP A NEW SYSTEM, TREAD SOFTLY. YOU MAY BE DISTURBING ANOTHER SYSTEM THAT IS ACTUALLY WORKING.

28. The Fail-Safe Theorem:
WHEN A FAIL-SAFE SYSTEM FAILS, IT FAILS BY FAILING TO FAIL SAFE.

29. COMPLEX SYSTEMS TEND TO PRODUCE COMPLEX RESPONSES (NOT SOLUTIONS) TO PROBLEMS.

30. GREAT ADVANCES ARE NOT PRODUCED BY SYSTEMS DESIGNED TO PRODUCE GREAT ADVANCES.

31. The Vector Theory of Systems:
SYSTEMS RUN BETTER WHEN DESIGNED TO RUN DOWNHILL.
Corollary:
SYSTEMS ALIGNED WITH HUMAN MOTIVA-TIONAL VECTORS WILL SOMETIMES WORK. SYSTEMS OPPOSING SUCH VECTORS WORK POORLY OR NOT AT ALL.

32. LOOSE SYSTEMS LAST LONGER AND WORK BETTER.
Corollary:
EFFICIENT SYSTEMS ARE DANGEROUS TO THEMSELVES AND TO OTHERS.

Advanced Systems Theory
The following four propositions, which appear to the author to be incapable of formal proof, are presented as Fundamental Postulates upon which the entire super-structure of General Systemantics, the Axioms in-cluded, is based. Additional nominations for this cate-gory are solicited (see Appendix III).
1. EVERYTHING IS A SYSTEM.
2. EVERYTHING IS PART OF A LARGER SYS-TEM.
3. THE UNIVERSE IS INFINITELY SYSTEMA-

TIZED, BOTH UPWARD (LARGER SYSTEMS) AND DOWNWARD (SMALLER SYSTEMS).

4. ALL SYSTEMS ARE INFINITELY COMPLEX. (The illusion of simplicity comes from focussing attention on one or a few variables.)

APPENDIX II:
FIRST BIENNIAL READERS'
SELF-EVALUATION QUIZ* FOR TESTING
MASTERY OF BASIC GENERAL
SYSTEMANTICS

This quiz consists of a series of brief examples illustrating basic principles of the operation of large systems. You are asked to read each example and then to indicate (in the space provided) as many as possible of the basic Systems-Axioms which apply. (Advanced students may indicate the Axioms by number rather than by name.)

1. You dial the telephone number of a friend in a nearby suburb. A recorded voice comes on the line, informing you that you have dialed incorrectly and instructing you to reread the directions in the front of the telephone book. Resisting the urge to answer back, you mutter to yourself: "Axioms number_____, _____, and _____, also_____ and _____."

2. The Titanic, designed to be unsinkable, had twenty-four bulkheads, each of which ran the full width of the ship. When she grazed an iceberg, however, the rent in the hull ran full length, breaching all twenty-five compartments. Axioms number _____, _____, and _____.

3. You are taking an examination in College Economics. The first question reads: Was President Franklin Roosevelt's Gold Policy a success or a failure? As a Systems-student, you immediately think of Axioms number _____, _____, and perhaps also _____.

*Open-book, of course.

4. On a bright April morning you receive a Christmas card in the mail. It is postmarked December 20, two years ago. Before handing it to you, the mailman demands two cents postage due, because the price of a stamp has gone up since it was mailed. Axioms number _____, _____, and _____.

5. A child psychiatrist, wishing to be both modern and efficient, as well as to gather research data on his practice, develops a questionnaire for parents to fill out. It includes questions on the nicknames, hobbies, and personal idiosyncrasies of relatives out to the level of third cousin. He presents the questionnaire to Mrs. Ept, whose son, Newton N., has been having trouble in school. When confronted with the questionnaire, Mrs. Ept refuses to fill it out, announces that the doctor is an idiot, and takes her child home.* The psychiatrist has failed to abide by Axioms number _____, _____, and _____.

6. A computerized study of funds managed by institutional investment counselors over the past thirty years demonstrates that such funds have grown (and shrunk) at precisely the rate predicted on the basis of *random* decisions to buy and sell stocks. Axioms number _____, _____, and _____.

7. Graduate schools train people for intellectual leadership by keeping them in the role of submissive students until middle age. Axioms number _____, _____, and _____.

8. Medical students, many of whom are destined to become family doctors, are trained in great centers of tertiary-level medical care, where common ailments are rare and rare entities are common. They learn to treat almost everything that they will never see again. They do not learn to treat what they will encounter every day. Axioms number _____, _____, and _____.

*The boy later became a Rhodes scholar.

Essay Questions (Advanced Students Only)

1. Discuss the impact of television on the design of municipal sewage systems.

2. The development of the Peruvian fishing industry may have resulted in less protein than before for the undernourished children of Peru. Explain.

3. Discuss, from the Systems-standpoint, the following statement: Prisons contain those who are *not deterred from crime* by the threat of imprisonment.

4. Explain:

(a) why no major organization has ever voluntarily disbanded itself;

(b) why major doctrinal advances are rarely made by the chief officer of a religion, philosophy, or political party;

(c) why company presidents rarely if ever introduce a major change in the function or product of the company.

APPENDIX III:
READERS' TEAR-OUT FEEDBACK SHEET

As indicated above, this catalog represents only a preliminary listing of the most basic and immediately obvious of the Systems-Axioms. You, the reader, may well be aware of one or more Systems-Axioms that have been omitted, or perhaps this work has stimulated you to think up some of your own.

Please use the space below to state it (them) in the briefest form commensurate with ready understandability. New Axioms thus acquired will be submitted to a panel of impartial judges (the author plus anyone nearby at the time), and the best of them will be juried in for inclusion in succeeding editions (if any) of this work. Here is your chance to achieve immortality, even if anonymously!

Axiom #1:

Axiom #2:

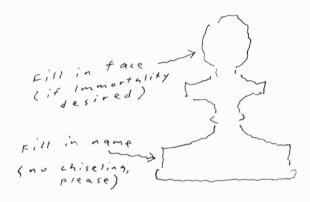

APPENDIX IV:
ANNUAL "AUNTY" AWARDS
FOR
SYSTEMS ANTICS OF THE YEAR
AND
WHOLE SYSTEMS CATALOG
OF OUTSTANDING EXAMPLES OF THE
OPERATION
OF THE LAWS OF GENERAL
SYSTEMANTICS

First Prize

First prize in the current cycle goes to the Nixon White House for a beautifully coordinated series of examples of the operation of Systems-Laws dating back to 1968 and even earlier. The Prize is awarded specifically for a truly classic demonstration of Axiom Number 19, the Newtonian Law of Systems-Inertia. When the Watergate story began to come out, the authors of the cover-up proceeded to try to cover up the cover-up, thereby demonstrating that:

A System Continues To Do Its Thing, Regardless
Of Circumstances.

Second Prize

Not quite in the same league as our First-prize winners, the runners-up proved that they are ready to provide keen competition to all contenders with the following specimen of System-behavior.

The Kennedy Foundation, established to *advance new ideas* in the field of medical ethics, announced that its first official act would be to fund a grandiose plan for *computerized retrieval of all the old, dead ideas* on the subject embalmed in the medical literature.

While demonstrating the truth of the old proverb (which is *not* a Systems-Axiom) that a Tool is no Wiser than its Wielder, the Foundation has also made it clear that Systems, alas, Never Quite Manage to Do What We Really Wanted—a touching reaffirmation of the ubiquitous Operational Fallacy, Axiom Number 8.

First Honorable Mention

The judges were unanimous in awarding First Honorable Mention jointly to the United States Coast Guard and the Canadian Environmental Protection Service for their Operation Preparedness, a proposal to study the effects of an oil spill upon the ecology of Lake Saint Clair (above Lake Erie) by actually dumping five hundred gallons of jet fuel into the lake. (Advanced students should find at least three applicable Axioms.)

Special Award

A Special Award goes to the designers, builders, and operators of supertankers, those gargantuan vessels that carry as much as half a million tons of oil around the tip of Africa to Europe and America. Supertankers exhibit many features of interest to the Systems-student. For example, they have a draft of up to sixty feet—too deep for most of the ports at which they call, and indeed too deep for safe navigation of the English Channel and the North Sea.

To save money, their operators deliberately send them into the wildest water on earth, some twenty miles off the Cape of Good Hope. Here they are battered by 80-foot waves. Too massive to ride with the waves, they take the full force of mountainous seas on their bows. But the captain, isolated on the bridge a thousand feet astern, cannot see the bow of his own ship. Even if he should suspect damage, he can do nothing, as there is no way to get forward in bad weather—there is no below-decks passage.

Supertankers are equipped with only one boiler

and one screw. If either fails, the ship drifts at the mercy of wind and wave. The one boiler provides electricity for lights, radio, and radar. This example of Bottleneck Design guarantees that the slightest malfunction can be amplified into a major disaster. If the boiler fails, all shipboard functions go dead within twenty minutes. An alarm system signals a malfunction, but does not indicate *where the problem is.*

But these features of supertankers, while interesting, have little fundamental significance for General Systemantics, since the defects of design and errors of operation are glaringly obvious. Simple greed is not, per se, a Systems-function. The *specific effect* for which the special Award is given is the following:

Supertankers exhibit an unexpected tendency to explode and sink on the high seas—not when loaded with oil, but when *returning empty* to their home port.

The cause is not well understood but may be related to spontaneous electrical discharges occurring in the oil-soaked atmosphere of the cavernous hold.

In the opinion of the judges, this is the year's best example of the Non-Additivity (Climax Design) Theorem.

APPENDIX V:
ANTICS ENTRY BLANK

The space below is set aside to provide an opportunity to Systems-students of every discipline to register their own contenders for the Annual "Aunty" Award Contest. The rules are simple: describe (as briefly as possible) your Horrible Example. Provide documentation or enclose the original report. Indicate which Axioms (in your opinion) are involved. Write in your name and address. Then mail this page to us.

Horrible Example:

Reference or other documentation:

Axioms involved:

Your name and address:

You may submit as many entries as you like, using for each a format similar to the above.

APPENDIX VI:
GLOSSARY

Anergy. The negative of energy. In biological systems, torpor. The amount of energy it would take to clean up some situation you don't like. Anergy resides within messy situations as energy resides within a coiled spring. A coiled spring is full of energy. When fully uncoiled, it is full of anergy.

Axiom, Axiomatic Method. The logical and necessary approach to developing the science of General Systemantics. The traditional approaches of *Observation* or of *Experiment* are clearly inadequate; the former because progress bogs down in impenetrable swamps of data, the latter because experiments upon systems invariably distort them beyond recognition.

The correct approach is to enunciate the Axioms from the start and then to show that they apply universally with only apparent exceptions. All that is necessary is to think very clearly at the most fundamental level and then to state one's clear thoughts in the briefest possible way. This saves endless bother. The result is an Axiom, and it will be found to be at least as accurate as any generalization achieved in more conventional ways. Euclid used the method to advantage; so can we. And everyone knows how successful it was in the hands of Descartes, who looked within himself and saw very clearly that he was thinking. The resulting Axiom: "I think; therefore I am" emerged with disarming ease and spontaneity.

Creativity, Scientific. The art of finding problems that can be solved (Warburg). In General Systemantic theory:

the art of recognizing simple systems. Often enough, the creative act consists of recognizing the simple system buried in the messy problem, i.e., of restating an existing problem in such form that it can be solved.

Obviously, systems cannot restate the problems they were designed to deal with, nor can a system recognize simple systems. Only people can do that. In fact, despite appearances to the contrary, systems cannot think at all.

Efficiency. Before one can estimate efficiency, one must first decide the *function* of the system. Since most large systems have multiple functions, many of which are not apparent on casual inspection, true efficiency is exceedingly difficult to estimate.

Efficiency Expert. Someone who thinks he knows what a system is or should be doing and who therefore feels he is in a position to pass judgment on how well the system is doing it. At best a nuisance, at worst a menace.

Evaluation. The process by which the System ascertains that the work it has done is genuinely good. Compare Genesis 1:31. Advanced Systems periodically review and evaluate their own evaluation procedures. This produces an infinite regression or incestuous process, but no one pays any attention to that.

Expert. A person who knows all the facts except those necessary to ensure the successful functioning of the system.

Function. In large systems, an intangible attribute not susceptible to easy definition. Often equivalent to what you think the system is doing, or whether you think it is doing it.

Garbage. A product of a system, for which no immediate use is apparent. The point to remember about garbage is that one system's garbage is another system's precious raw material. Clearly, garbage, like beauty, is in the eye of the beholder, and the amount of garbage in the world is a function of the viewer's imagination and ingenuity.

Goal. What you want the system to do. The important

thing to remember is that the designed-in function of the system is probably something very different.

Objective. A lesser goal, greater than a Sub-objective but not sufficiently grand to be an End-in-itself. A logical fraction of a Total Goal. Example: If the Goal is to resolve the structure of DNA, an Objective might be to resolve the left end of the molecule. A separate team of workers would logically be assigned to that Objective.

Problem. When spelled with a capital letter, the Problem is a statement of how the System conceptualizes the real-world problem. Real-world problems cannot be solved by Systems, because the function of the System is limited to an already existing conceptualization and real-world problems are resolved by radical innovation, not by new combinations of old ideas. Compare the "ether" problem that plagued nineteenth-century physics or the "phlogiston" problem that bedevilled eighteenth-century chemistry.

System. "A set of parts coordinated to accomplish a set of goals." An eminent Systems-student,* after offering this as a definition that all will agree upon, then devotes an entire chapter to the question, "What is a system?" Clearly, a dictionary definition is not enough. His own definition of a "systems approach" to a problem is "the whole set of subsystems and their plans and measures of performance."

May we conclude that a System is a *set of subsystems* coordinated in some way for the achievement of some purpose? If so, what is a subsystem? Are Systems and Subsystems infinitely divisible both upward and downward?

We leave this deep metaphysical question for the reader to ponder at his leisure.

Systems-Theory. There are some who assert that General Systemantics is a spoof of a serious scientific subject

---

*Churchman, *op. cit.,* p. 29.

called General System Theory.* Devotees of General System Theory attribute the founding of their science to Professor Ludwig von Bertalanffy,** who noted, in the early
decades of this century, that scientists had overlooked the
establishment of a Science of Anything and Everything
and who, with Teutonic thoroughness, made up the oversight.***

*While denying, on principle, that such is the case, we admit to
certain parallelisms in development of the two fields.
**The name is genuine.
***The twentieth century has seen a good deal of this kind of
intellectual tidying-up. In biology, Professor Hans Selye under-
took to elucidate the pathology of the disease called Life. In
Philosophy, Count Korzybski founded the Science of Meanings.
Modern Physics seems to be devoted to the investigation of Less
and Less; and in Mathematics, vigorous attacks are being made
upon the Knowledge of Absolutely Nothing.

BIBLIOGRAPHY

Arrowsmith, William. "The Shame of the Graduate Schools," *Harper's Magazine,* March 1966, pp.52–59.

Atkinson, B.M., Jr., and Whitney Darrow, Jr. *What Dr. Spock Didn't Tell Us.* New York: Dell, 1958.

Bennis, Warren. "Who Sank the Yellow Submarine?" *Psychology Today,* November 1972, pp. 112–120.

Berne, Eric. *Games People Play.* New York: Ballantine Books, Inc., 1964.

Bertalanffy, Ludwig von. *General Systems Theory.* New York: Braziller, 1968.

Brewer, Garry. *Politicians, Bureaucrats, and the Consultant. A Critique of Urban Problem Solving.* New York: Basic Books, 1973.

Carroll, Lewis. *Alice's Adventures in Wonderland* and *Through the Looking Glass* in One Volume. Illustrated by John Tenniel. New York: Heritage Press, 1941.

Churchman, C. West. *The Systems Approach.* New York: Dell, 1968.

Fiedler, Fred E. "The Trouble with Leadership Training Is That It Doesn't Train Leaders." *Psychology Today,* February 1973, p. 23.

Harris, Thomas A. *I'm OK—You're OK.* New York: Harper and Row, 1967.

Herbert, A.P. *What A Word!* New York: Doubleday, 1936.

Janis, Irving L. "Groupthink." *Yale Alumni Magazine,* January 1973, pp.16–19.

Korzybski, Alfred. *Outline of General Semantics.* In General Semantics. Papers from the First American Congress for General Semantics. Organized by Joseph C. Trainor and held at Ellensburg, Washington, March 1 and 2, 1935. (Central Washington College of Education). Collected and Arranged by Hansell Baugh. Distributed by Arrow Editions, 444 Madison Avenue, New York, 1938. Copyright 1938 by the Institute of General Semantics, 1330 East 56th St., Chicago, Illinois.

Laotzu. *The Way of Life According to Laotzu.* An American Version by Witter Bynner. New York: John Day, 1944.

McLuhan, Marshall. *Understanding Media: The Extensions of Man.* New York: New American Library, 1964.

Mendelssohn, Kurt. "A Scientist Looks at the Pyramids," *American Scientist,* March-April 1971, pp. 210–220.

Orwell, George. *Nineteen Eighty-Four.* New York: Harcourt Brace Jovanovich, 1949.

Parkinson, C. Northcote. *Parkinson's Law and Other Studies in Administration.* Boston: Houghton Mifflin, 1957.

Peter, Lawrence J., and Hull, Raymond. *The Peter Principle.* New York: Bantam, 1970.

Potter, Stephen. *One-Upmanship.* New York: Holt, 1951.

Rosenhan, D.L. "On Being Sane in Insane Places," *Science,* January 1973, pp. 250–258.

Rostow, Eugene V. "In Defense of the Ivory Tower." *Yale Alumni Magazine,* June 1972, pp. 5–7.

Russell, Bertrand. *Freedom Versus Organization.* New York: Norton, 1962.

Selye, Hans. *The Stress of Life.* New York: McGraw-Hill, 1956.

Solzhenitsyn, Alexander. *We Never Make Mistakes.* New York: Norton, 1963.

Townsend, Robert. *Up the Organization.* Greenwich, Connecticut: Fawcett, 1970.

Whyte, Jr., William H. *The Organization Man.* New York: Doubleday, 1956.

Yutang, Lin. *The Importance of Living.* New York: John Day, 1937.

DATE DUE
